MAGICAL DISPLAY

The Art of Photomicrography

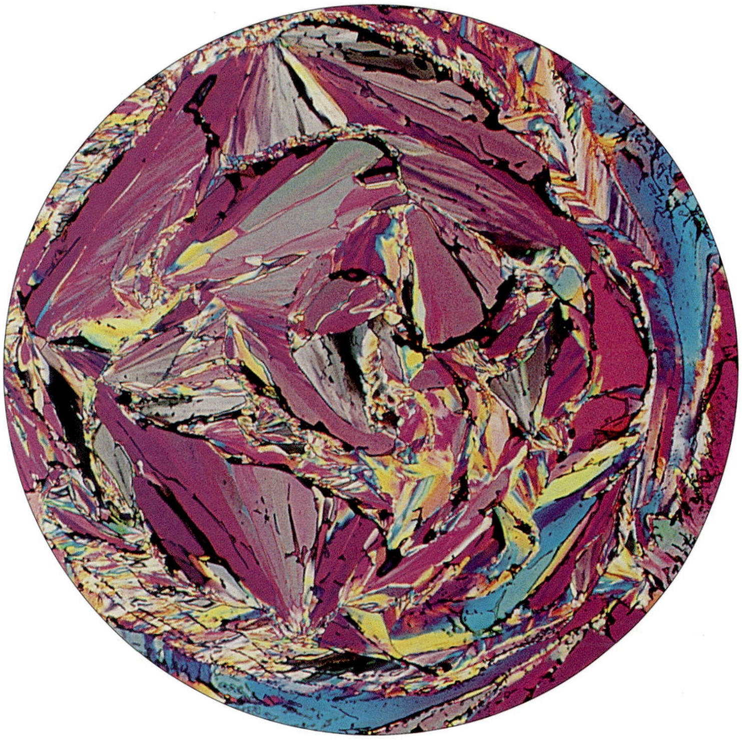

MICHAEL W. DAVIDSON

AMBER LOTUS

First published 1993

Institute of Molecular Biophysics (IMB)
Florida State University, Tallahassee, Florida

For information write to:

Amber Lotus, 1241 21st Street, Oakland, CA 94607
Amber Lotus is a division of Dharma Enterprises.

Printed in USA on recycled paper by Dharma Enterprises.

ISBN: 0-945798-83-0

CONTENTS

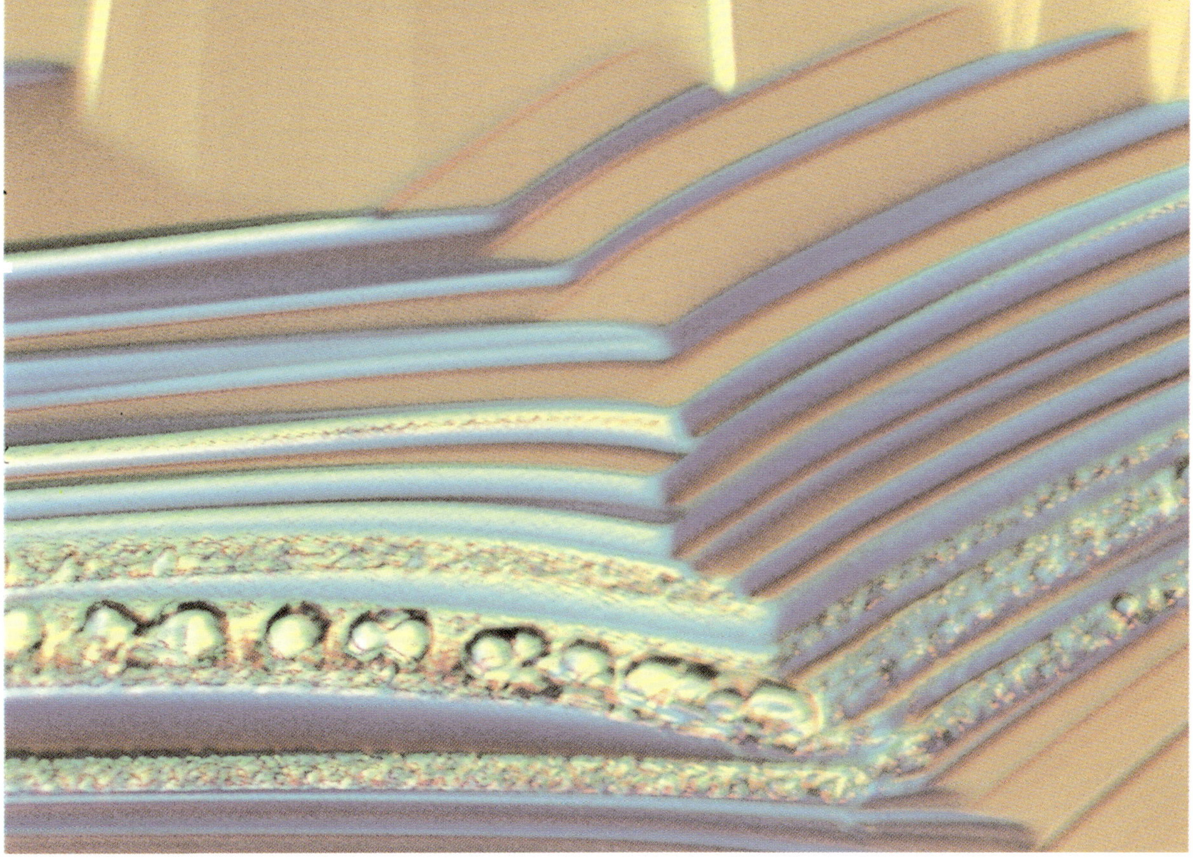

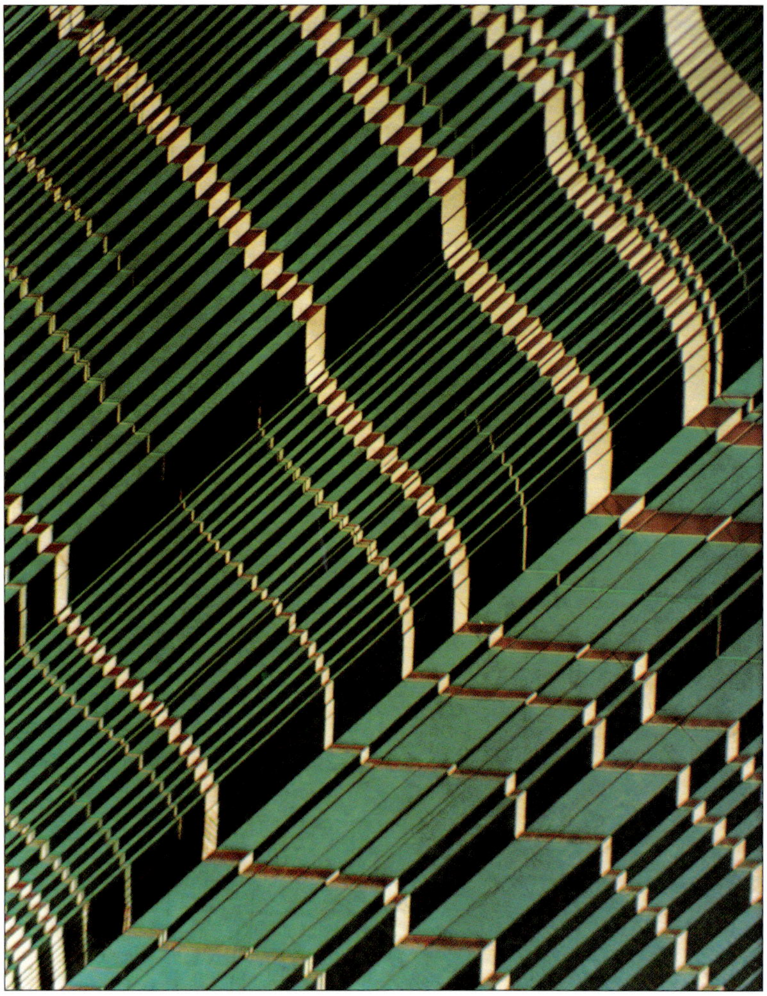

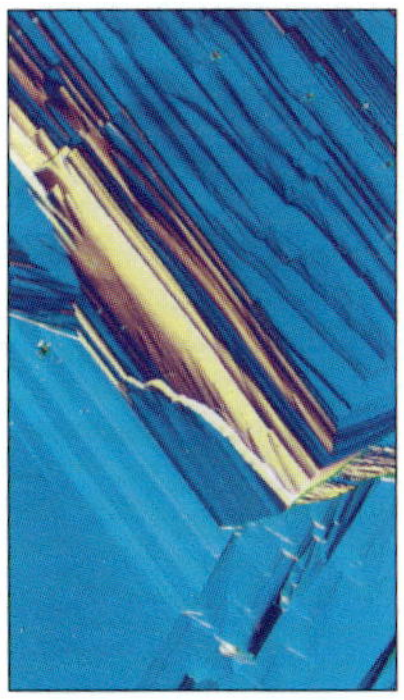

INTRODUCTION

Practically everyone has, at one time or another, viewed the world through an optical microscope. For most people, this experience occurs during biology training in high school or college.

Photography through the microscope, or more commonly, *photomicrography*, has long been a useful tool to scientists. The biological and medical sciences have, for many years, relied heavily on microscopy to solve problems relating to the overall morphological features of specimens as well as a quantitative tool for recording specific optical features and data. In this respect, the optical microscope has proven useful in countless investigations into the mysteries of life.

More recently, microscopy has enjoyed an explosive growth as a tool in the physical and materials sciences as well as the semiconductor industry, due to the need to observe surface features of new high-tech materials and integrated circuits.

In classical biological microscopy, thin specimens are prepared and the light is passed or transmitted through the sample, focused with the magnifying lens (commonly referred to as an objective) and into the eyepieces of the microscope. For observing the surface of integrated circuits (that comprise the internal workings of modern computers) light is reflected from the surface of a sample and into the microscope objective. In scientific nomenclature, transmitted and reflected light microscopy are known as *diascopic* and *episcopic* illuminated microscopy, respectively. The photomicrographs illustrated here are from both transmitted and reflected optical microscopic scientific investigations.

One of the biggest problems in microscopy is the poor contrast produced when light is passed through very thin specimens or reflected from surfaces with a high degree of reflectivity. To circumvent this lack of contrast, various optical

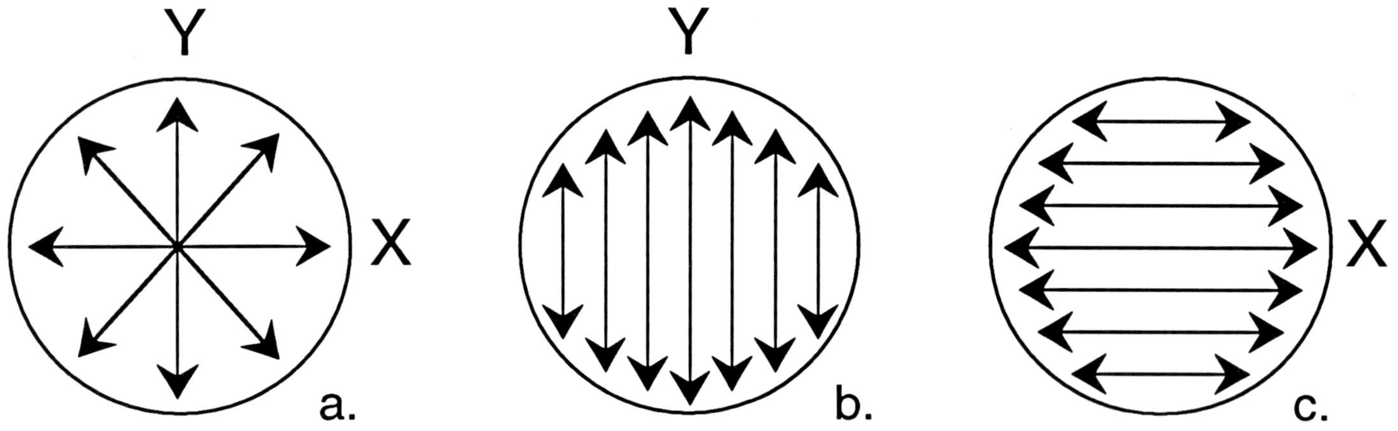

Figure 1. Polarization of light. (a) White, unpolarized light; (b) Plane-polarized light in the vertical direction; (c) Plane-polarized light in the horizontal direction.

"tricks" have been perfected by scientists to increase contrast and to provide color variations in specimens. The assortment of techniques in the microscopist's bag include: polarized light, phase contrast imaging, differential interference contrast, florescence illumination, darkfield illumination, Rheinberg illumination, Hoffman modulated contrast, and the use of various gelatin optical filters. A thorough discussion of these techniques is well beyond the scope of this book, therefore I will concentrate only on the methodology that was used to compose the photomicrographs reproduced here. References at the end of this chapter should serve to provide more details of microscopy and photomicrography to interested readers.

A majority of my photomicrographs are taken with optically polarized transmitted light. This means that the visible white light is first passed through a polarizer so that all of the light rays are oriented in a single direction. Microscope polarizers are similar to those used on cameras and sunglasses to reduce glare. Figure 1 illustrates what happens to white light when it is passed through a polarizer. Imagine that you are looking directly into a beam of white light generated by a flashlight. Figure 1a illustrates the vibrational planes of the many optical components that vibrate in all directions. When this light is passed through a polarizer limited to vibrational planes in a single direction, the emerging light will all be *plane-polarized* in that direction (X in Figure 1b and Y in Figure 1c).

The polarized light next passes through the sample and into another polarizer that polarizes light in a plane that is perpendicular (or rotated 90°) with respect to the first polarizer (See Figure 2). The second polarizer is often termed an analyzer in polarized microscopy. The resulting setup is termed *crossed polarization*, meaning that without a sample between the polarizers, the viewfield in the microscope is black with no light coming through. For observation, a sample (usually a recrystallized chemical on a microscope slide) is placed on the microscope stage between the crossed polarizers. Samples that have a property called *birefringence* will be able to twist some of the light so that it can pass through the second polarizer and be imaged in the eyepieces. Birefringence is a complicated and very detailed concept. Thus, to make it easier to understand, we can imagine that the recrystallized chemicals in the sample act in a manner similar to miniature prisms to refract, twist, and bend the light rays so they can pass through the second polarizer.

Many different kinds of samples are birefringent, including the liquid crystals found in many portable computer displays, single crystals of pure chemicals, polymers (plastics), and the minerals found in thin sections of rock samples. Often, birefringent samples display a bright spectrum of colors when observed through a polarizing microscope. These beautiful colors are not the ones we generally equate with those observed in ordinary daylight, but they are *interference* colors formed in a sequence according to the path difference of the light beam through the birefringent material. A similar spectrum of colors is seen on soap bubbles or on a thin film of oil or gasoline on water. If a particular crystal formation refracts only green

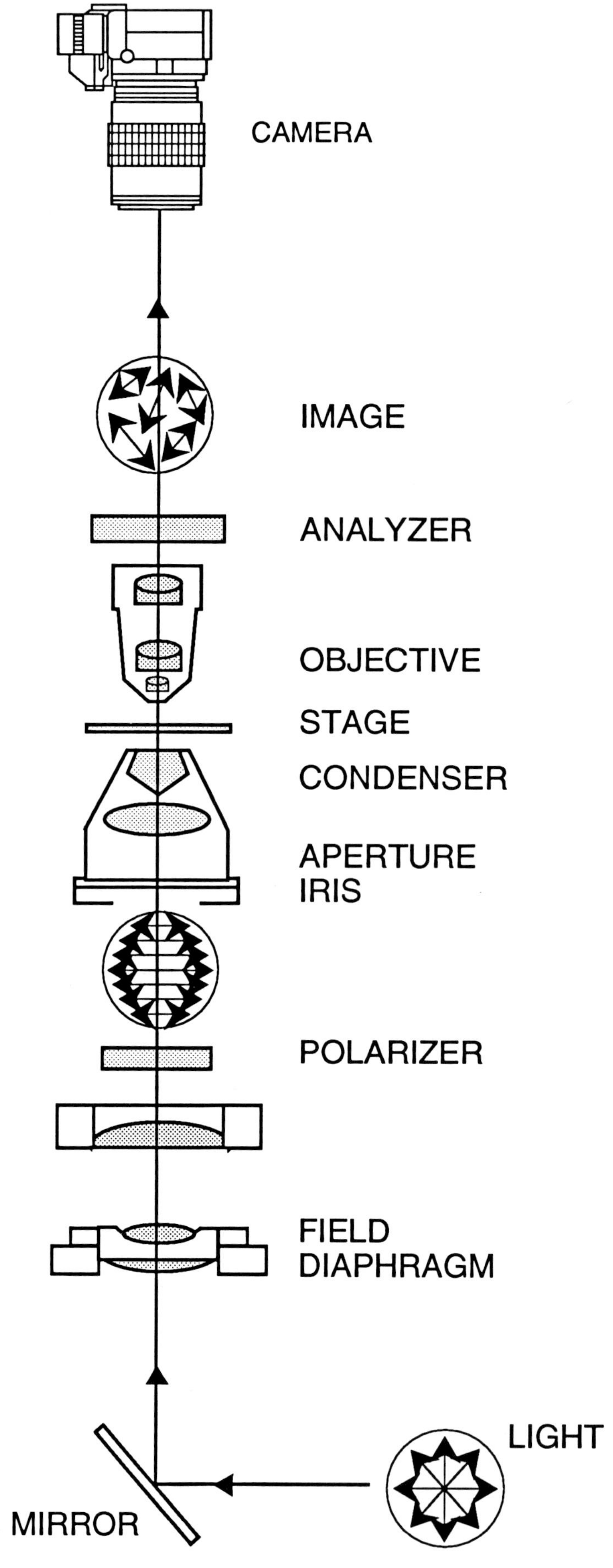
CAMERA
IMAGE
ANALYZER
OBJECTIVE
STAGE
CONDENSER
APERTURE
IRIS
POLARIZER
FIELD
DIAPHRAGM
LIGHT
MIRROR

Figure 2

light, for example, then the crystal will appear green in the microscope eyepieces. In reality, crystal formations are usually a mixture of many different crystallization patterns and give rise to practically the entire visible light color spectrum. This is the primary reason that the images in this book are so colorful.

Some of the specimens that we examine by microscopy are opaque and will not allow the passage of light necessary for transmitted polarized microscopy. In these cases, we use a technique termed reflected Differential Interference Contrast (DIC) microscopy where light is reflected off the surface of the sample and back into the microscope objective. Very smooth and shiny surfaces often show little surface detail due to the high reflectivity, and by adding DIC, we can enhance the quality of the photomicrographs and give them a 3-dimensional effect. This is especially useful for imaging uneven surfaces as illustrated in Figure 4b. The range of subjects that are best observed with reflected DIC microscopy include: integrated circuits, metals, ores, ceramics, and other high-tech materials such as composites, allomers, semiconductors, superconductors, and alloys.

Figure 5 depicts a schematic representation of reflected DIC microscopy. White light emitted by a tungsten-halide source is first passed through a polarizer, much like the situation in transmitted microscopy. Next, the polarized light is passed through a first order red compensator which is rotatable to vary color combinations. When deflected by a mirror, the light passes through a modified *Wollaston* prism which splits the light into two component beams, termed an object beam and a reference beam. The light is then directed through the objective and reflected from the sample surface back through the objective where it is recombined again by passing through the Wollaston prism and focused by the tube lens. After passing through an analyzer that serves the same purpose as a transmitted microscopy analyzer, the light is then focused onto film in a camera. DIC gives reflected microscopy subjects a 3-

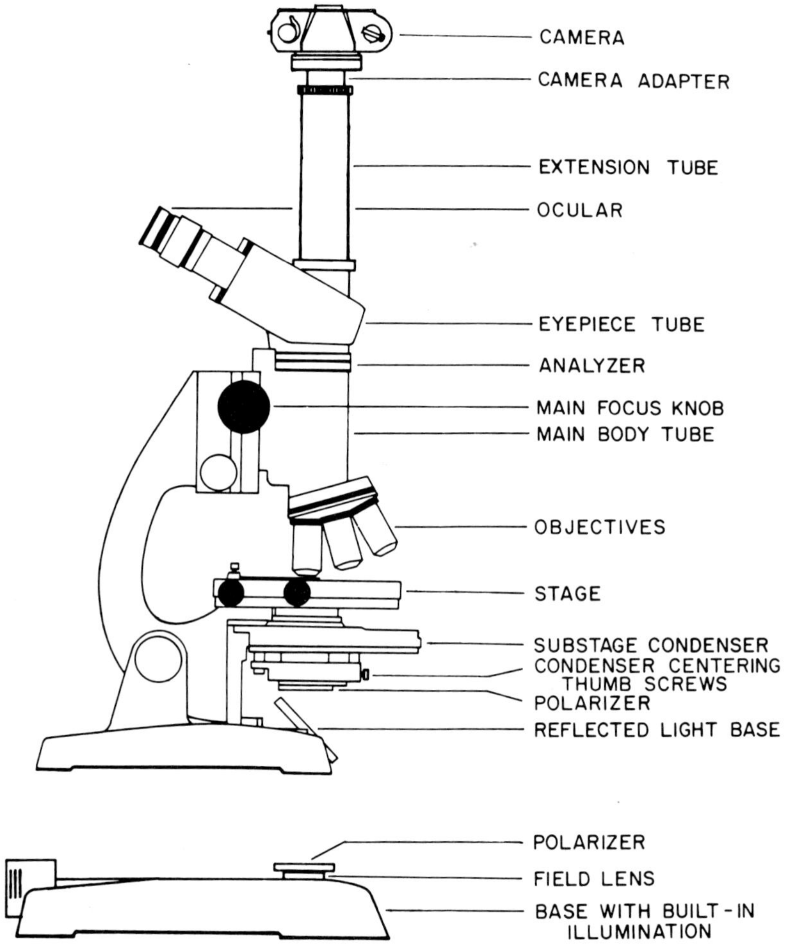

Figure 3. Optical transmitted polarized light microscope

dimensional effect that will display varying colors as the first order compensator is rotated.

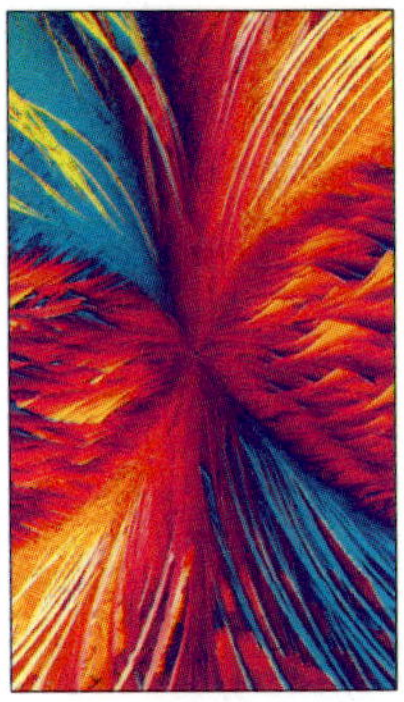

HOW TO SET UP A MICROSCOPE

Any brightfield microscope can easily be converted for use with polarizing elements. Two polarizers are needed to convert the microscope for use with polarized light. Many science supply houses and distributors offer an excellent selection of polarizing materials at somewhat low cost. When purchasing polarizing elements, remember to select materials that are as close to a neutral grey as possible. Avoid materials that are amber or green in color. These off-color materials will cause color shifts that must be corrected. The polarizer responsible for polarizing the light emitted from the microscope light source is placed either directly on the field lens or can be taped (with electrical tape) to the substage condenser (refer to Figure 3 for a detailed microscope description). On microscopes with an internally directed lamp, it is sufficient to place a polarizer onto the field lens as is illustrated in Figure 3. This polarizer should be large enough to cover the lens completely. Low cost polarizers designed for camera lenses are sufficient for this purpose. But, if an external light source is reflected into the substage condenser through a mirror, it becomes necessary to tape a polarizer onto the bottom of the condenser (Figure 3).

The second polarizer (which is termed the analyzer, as described above) is inserted inside the body of the microscope between the main body tube and the eyepiece tube. There is usually a lens mount covering the top body tube and the analyzer can be taped or simply rested directly on this mount. Because of the restricted space within the main body tube, the analyzer is limited in physical size to 1-3 cm in diameter. An analyzer of this size can be obtained by cutting a small section from a sheet of polarizing material or by purchasing a small polarizer from a dealer. The analyzer and upper main body lens should be

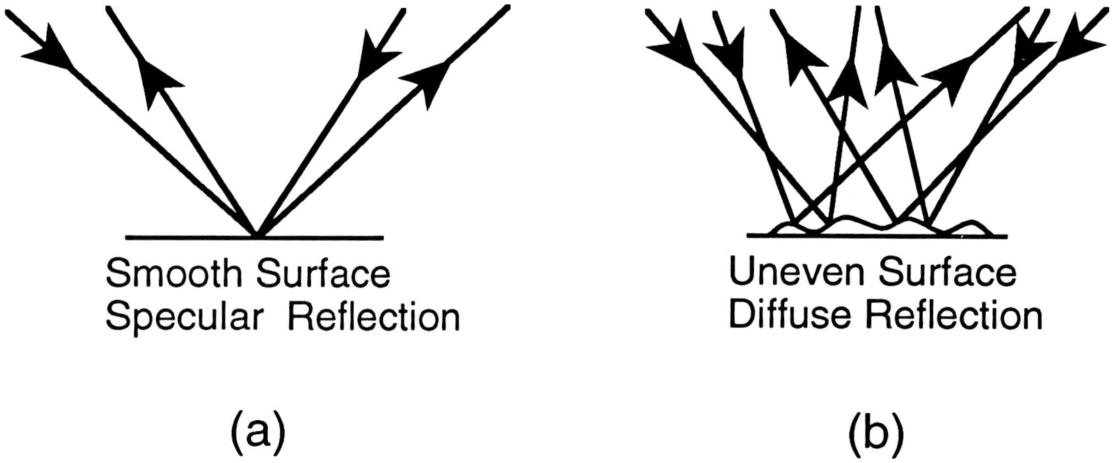

Figure 4. Reflected light from different surfaces. (a) Specular reflection from a mirror-like surface; (b) Diffuse reflection from an uneven surface.

thoroughly cleaned to eliminate dust and fingerprints before reassembly of the microscope. After both polarizers are in place, activate the microscope light source, place a specimen on the stage, and view directly into the oculars (eyepieces). Next, bring the image into focus and rotate the polarizer on the microscope base until the viewfield becomes very dark (maximum extinction). At this point, the polarization direction is perpendicular between the polarizer and analyzer and you then have crossed polarizers as described above. Some microscope manufacturers offer a low budget polarization kit ($150 to $300) that is easily user-installed. It is advisable to contact your microscope distributor or manufacturer concerning the availability of these items if your budget is healthy.

Quantification of optical birefringence can be performed by adding compensators, also known as tint, retardation, and λ plates, between the sample and the analyzer. These plates are fabricated from slabs of birefringent material (such as quartz, mica, and gypsum) that are cut in specially oriented crystallographic directions. In this manner, the direction of the fast and slow vibration components are known with respect to the axis of the crossed polarizers. These plates are usually inserted into a slot in the microscope near the analyzer so that the slow axis of the plate is situated at a 45° angle to the polarizer. The plates will retard the light by either a fixed or variable, but known, amount. The colors seen in the microscope when a compensator is inserted into the light path will be changed depending on whether the sample is in an additive or subtractive position with respect to the compensator. The compensators are sometimes useful to aid in photomicrography of very weakly birefringence samples, due to a change in background color from black to gray or magenta to red, depending on the wavelength retardation of the compensator plate.

Among the useful methods of enhancing sample contrast are the interference methods where contrast augmentation is similar to that in phase contrast microscopy with

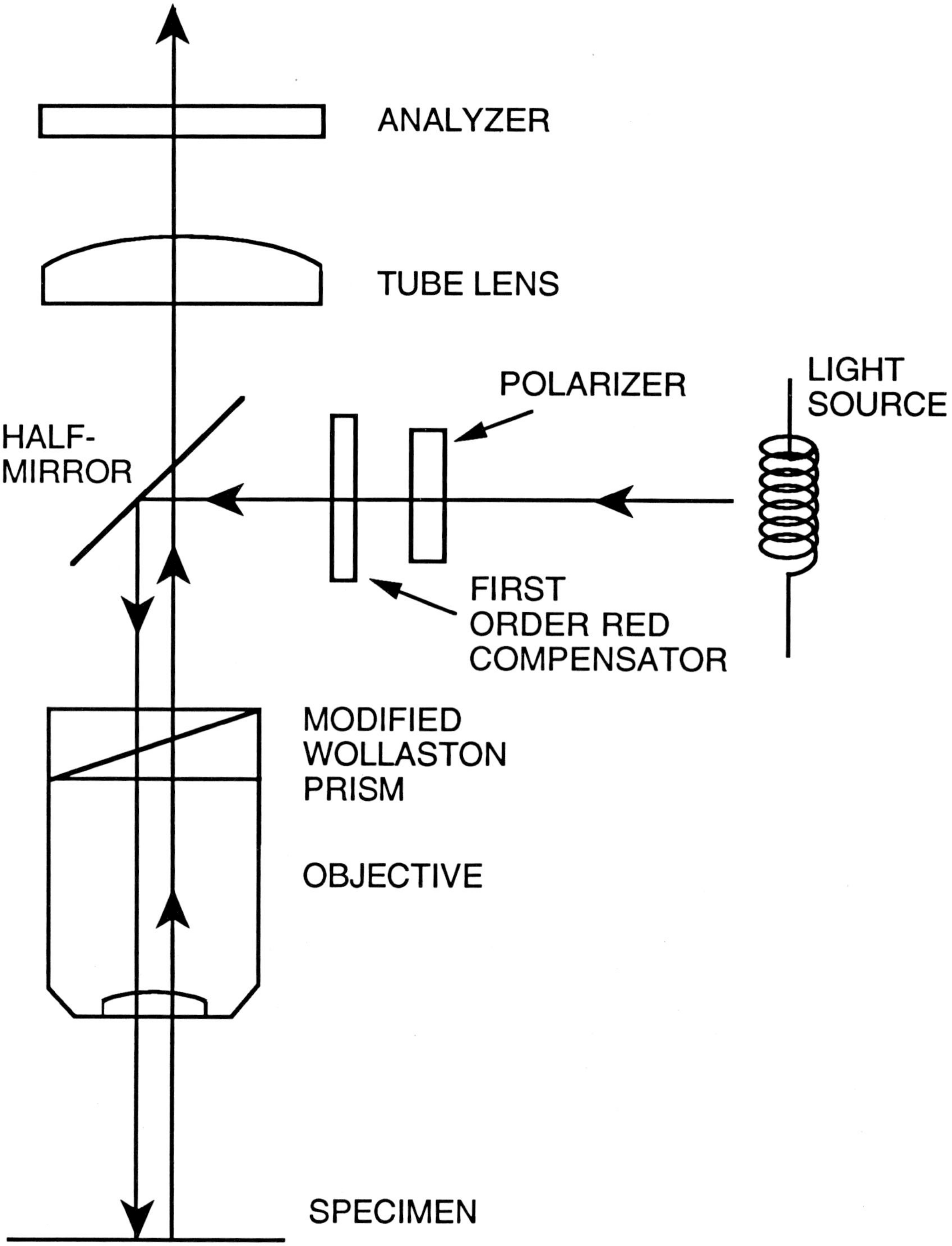

Figure 5. Reflected Differential Interference Contrast (DIC) Microscopy.

the elimination of edge "halo" effects. Differential Interference Contrast (DIC) microscopy requires expensive objectives equipped with a special set of prisms, termed Wollaston prisms, that are optically matched to each objective. A relatively low-cost substitute for the interference contrast effect generated by the Wollaston prisms involves adapting a polarized light microscope with *Savart plates* serving in place of the polarizers. Savart plates are flat plates of a suitable quartz crystal cut and polished to produce the necessary divided interference beams of light in order to generate the three- dimensional effect of interference contrast. These plates are commonly available from optical supply houses.

Another method of contrast enhancement was developed by microscopist Julius Rheinberg in 1896. Rheinberg illumination is often referred to as *optical staining*, and lends an exciting spectrum of color and highlights to conventional microscopy techniques. It is especially useful for amorphous and unstained samples which lack sufficient detail for successful photomicrography. Rheinberg illumination is similar to conventional darkfield illumination except that the central opaque light stop at the base of the substage condenser (Figure 3) is substituted for a colored central stop. A sample viewed with the colored central stop will appear white on a background the color of the stop. Addition of a colored transparent annular filter will add color to the sample. A variety of materials can be employed for constructing the colored stops and annular rings. Colored cellulose acetate or gelatin filters are probably the most convenient, and many graphics supply houses sell, or sometimes give away, small sample books of these materials. By experimenting with a variety of different colored stops and annular filters, a wide spectrum of different images can be obtained.

The renovations described above apply only to transmitted light microscopy where polarized visible light passes directly through the sample. In reflected light microscopy, the light beam is reflected from the surface of the sample and scattered into the microscope objective. To avoid investing in expensive reflected light attachments, oblique illumination from an external light source can be substituted to produce the reflected light effect. A high-intensity light source such as a fiber optics lamp (available for under $200) provides an excellent substitute. To achieve a semi-polarized light effect, the analyzer can be left in place and polarizers can be attached (by tape) to the external light source. Before final attachment, the polarizers should be rotated as described above to produce maximum extinction. Unpolarized room light, that will unavoidably be reflected into the objective, will slightly diminish the total amount of light extinction. This method of microscopy is particularly well suited for examination of integrated circuits and other specimens which possess intricate surface details.

Common biological stereo-binocular (dissecting) microscopes are also useful for reflected light examination of materials science and geological rock and mineral samples, especially for photomicrography of rocks and crystals. These microscopes are often useful in the brightfield mode, but can also be operated with polarized light to obtain details of the surface structure of birefringent materials. To adapt a stereo microscope for polarized light, tape a large polarizer (such as a model designed to mount onto a camera lens) onto the lower portion of the main body tube. Insure that the polarizer completely covers the lens mounted in this lower portion of the tube. Next, adapt a polarizer to the external light source. If a fiber optics illumination device is used, simply tape the polarizer over the end of the fiber outlet tube. Fiber optics sources that have a beam splitter can be obtained to divide the output light into two tubes. These sources are ideal for illumination of samples with reflected light because a sample can be illuminated from several directions to eliminate a shadow effect. In this case, be sure to cover both fiber outlet tubes with a polarizer. First, place a mirror on the stage to reflect light directly into the objective and rotate the polarizers on the fiber outlet tubes until maximum extinction is reached. It is usually easier to block the light from one tube and adjust the other for maximum extinction, repeating the process again for the blocked tube.

Attaching a camera to the microscope is the very last step. Microscope viewing heads come in three varieties: monocular (one eyepiece), binocular (two eyepieces), and trinocular (two eyepieces and a photography tube). A camera can be adapted to each of these viewing heads. Commercially available after-market camera adapters usually are attached to one of the viewing tubes with a thumbscrew and adjusted to be parfocal with the eyepieces by sliding the adapter up or down on the viewing tube. A simple camera back is sufficient for photomicrography because the camera is required only to store, expose, and advance the film. The microscope itself acts as a camera lens. After a camera back has been adapted to a microscope, you should not rely on exposure values computed by in-camera exposure monitors. It is best to *bracket* exposures over several f-stops, as discussed below, to get a handle on exposure times as is discussed in the section on photomicrography.

For optimal photomicrography, it is important to insure that your microscope is aligned to produce an even illumination across the viewfield. Information on microscope alignment is available in owners manuals or in textbooks dealing with microscopy.

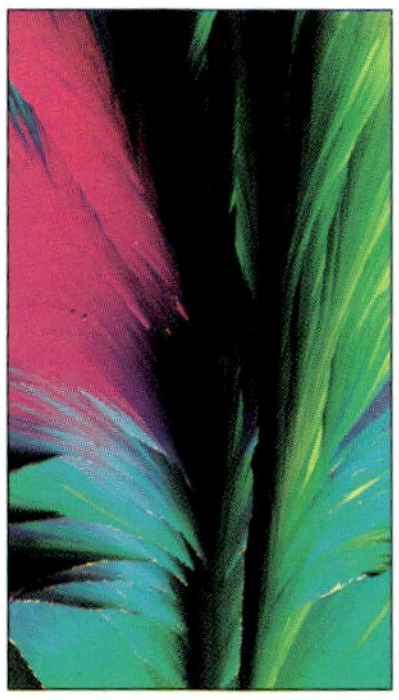

PHOTO-MICROGRAPHY

Recording images seen in the microscope onto photographic film allows scientists to produce a "hard copy" for research records. In an educational environment, classical photography assignments can be coupled with science microscopy studies to provide a multidisciplinary program in photomicrography.

Photomicrography encompasses the techniques of both black

and white and color photography. Black and white film processing is substantially lower in cost than color film, if processing is done in-house. Unfortunately, many commercial film processors no longer offer black and white processing services or charge exorbitant prices for this service. If budget restrictions force the use of black and white photomicrography, it is advisable to invest in darkroom equipment so students and research assistants can develop and print their own photomicrographs.

A green gelatin or interference filter should be inserted into the microscope lightpath between the light source and the first polarizer (see Figure 3) for black and white photomicrography. With built-in illumination, the filter can be placed on the field lens prior to the placing of the polarizer. On microscopes that use an external light source, the filter can be taped below the polarizer on the substage condenser. I recommend the use of Kodak Technical Pan film (or it's equivalent) with HC-110 developer for crisp images with excellent contrast and resolution. This film provides intricate detail imaging and reduces the continuous tonal tendencies of the popular black and white films. Printing can be done on Kodak Polycontrast (or again, it's equivalent) paper with Dektol developer. After processing a roll of film, carefully cut the negatives into sections of 5 frames each and store these in specially made polyethylene storage sheets. Contact prints can be made by placing a sheet of negatives directly onto an 8" x 10" piece of Polycontrast paper and exposing for a few seconds with the enlarger lens aperture opened to the widest f-stop. Contact sheets are an ideal way of cataloging data and they provide a compact method for storing or sorting through many images. In case an enlargement is needed, you may simply remove the appropriate negative strip from the protective sheet and make your prints with an enlarger.

In some instances black and white transparencies are needed for projection during lectures and at seminars. If these are desired, it will be necessary to use a direct-positive reversal processing kit to prepare the transparencies.

Color photomicrography is considerably more complicated than black and white photomicrography because color film emulsions are color balanced for a particular spectrum of light. The term *color temperature* refers to the wavelength spectrum emitted by a particular light source. For instance, films intended to be used outside in ordinary daylight or under fluorescent lighting are balanced during manufacture for a color temperature of 5500°K while films made for indoor tungsten light bulb use are balanced for a color temperature of 3200°K. The majority of microscopes use a tungsten-halide bulb as a light source. These bulbs emit a wavelength spectrum centered in the 3200°K color temperature region. Therefore, films color-balanced for this type of illumination will produce the best results. Using daylight balanced films under tungsten illumination will shift all color tones towards a decidedly yellow cast. Likewise, using tungsten balanced films under daylight illumination will shift color tones towards a bluer cast. All major film manufacturers have one or several 3200°K films available in 35 mm transparency format. Transparency film is preferable to color negative film for several reasons. Most importantly, all color negative films are color balanced for 5500°K and must be manipulated during printing to avoid the yellow cast mentioned above. Most photo processors can not or will not produce satisfactory results with photomicrographs on color negative film. Also, the contrast and color saturation in transparency film cannot be equalled by color negative film. Finally, color transparencies are easier to label, store, and catalog and they can be projected at seminars.

With a 20 to 100 watt tungsten-halide bulb in your microscope, exposure times are usually very short and allow the use of slow films such as Ektachrome 50 or Fujichrome 64T. Using slow films reduces the grain size in photomicrographs and leads to higher quality enlargements. If tungsten balanced films are not available, a Kodak 80A or equivalent filter can be inserted into the light-path between the light source and the first polarizer to allow the use of daylight balanced films with minimal color shift. If this filter is used, exposure times must be increased 1-3 f-steps to allow for a reduction in light intensity. Recently, Polaroid introduced a slow speed (ISO 40) color transparency film, designated Polachrome HC, with high contrast that is ideal for photomicrography. This film produces superb contrast and color saturation and can be user-processed in only 2 minutes with a low cost Polaroid processor.

When photographing new samples or after making changes to the microscope (such as installation of polarizers), the new exposure characteristics should be determined on a test roll of film. Bracket several exposures of the same viewfield at least one and preferably two f-steps over and under previous exposure times. This will assure at least one or several good exposures and will yield exposure time information that is useful for photomicrography of future samples.

The best film, in my opinion, is Fujichrome 64T, a highly color-saturated E-6 transparency film with excellent contrast. Recently, a new emulsion of this film was introduced that is designed to allow *push* processing with little reduction in image quality. Push processing is a method developed to increase contrast (inherently low in photomicrographs) and color saturation. This is done by underexposing the film 1 to 2 f-steps and increasing the process time in the first developer during the E-6 processing session.

The complicated E-6 process can be done in-house if the proper equipment is available. There are 7 steps to this process and the temperature must be carefully controlled to a constant 38.3°C ± 0.1°C. This is difficult to do without having a constant-temperature water bath dedicated to the task. A cheaper and easier alternative to the E-6 process is marketed by Kodak in the form of a "hobby pack" that contains the essential E-6 chemicals combined and packaged into 4 individual solutions. The hobby pack is much easier to use and has a far wider temperature latitude, although by following the temperature restrictions of the E-6 process, optimal results are obtained. I have found that identical, and in some cases superior results can be produced in the laboratory with this simplified process if a few rules are followed. The first developer is the most critical part of the process and it is essential that the temperature be held to within the specified limits. Contrast can be increased by addition of 3.0 grams of phenidone (commercially available at photography chemical distributors) per liter of first developer. The additional contrast obtained from addition of the phenidone dramatically helps the images to reproduce what is seen in the microscope oculars. In many instances, the processed film has either an overall yellowish-green or magenta cast after drying. This can be eliminated by adjusting the pH of the color developer. The optimum pH given by Kodak is 11.1, but by increasing the pH by 0.2-0.7 units any extra magenta can be subtracted, or by reducing the pH by the same amount, the yellowish-green cast can be eliminated. These hobby kits can be used to develop 10 rolls of film with 24 exposures in the magazines. As the solutions age, extended processing times are necessary to achieve the desired effects. Details on the time increases are given in an instruction booklet supplied with the kit. It is probably better to add 1–2 minutes to the color developer time given and add about 5 minutes to the bleach/fixer time to insure adequate bleaching of unwanted dyes.

By extending the time in the first developer by approximately 30%, you can push the process one f-step. This is tricky and it is advisable to experiment with non-valuable film during the initial setup for push processing. Too much time in the first developer will definitely lead to an increase in grain size and the transparencies will lose a significant portion of their overall density. Films that are push processed should be underexposed by one or several f-steps. It is a good idea to bracket a set of photographs in the same viewfield to determine the precise number of f-steps to underexpose the film. Push processing will increase contrast and also color saturation at the cost of a small sacrifice of the overall film density. Too much underexposure will leave the transparencies dark and dense. By monitoring a bracket composed of exposures at every half or third of an f-step, optimum underexposure can be determined.

For publication, it is sometimes necessary to obtain color prints made from photomicrograph transparencies. This can be done through commercial processors at a reasonable cost, however the quality is generally poor. Ilford produces a direct positive paper, called Cibachrome, which can be processed at room temperature in an inexpensive roller drum. Although it is rather expensive, Cibachrome produces prints that have superb contrast and color saturation. The process has only three steps: a developer, a bleach, and a fixer, and can be completed in 15 minutes. The roller drums handle 8"x 10", 11"x 14", and 16"x 20" prints. I usually purchase only the 11"x 14" paper and cut each piece into 4 smaller 5"x 7" sections. Four of the smaller prints can be exposed and loaded into an 11"x 14" roller drum and they can be processed simultaneously.

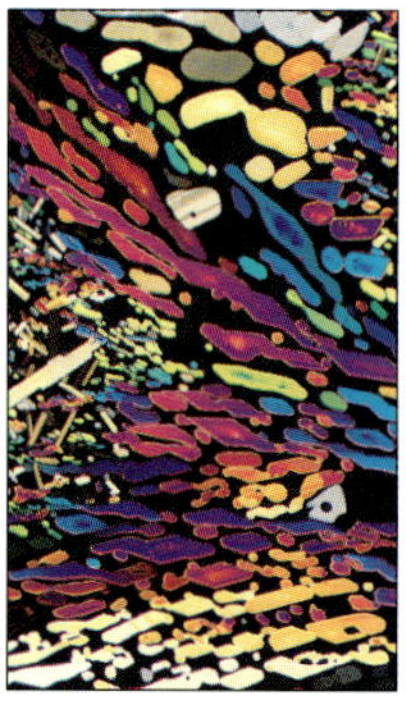

PREPARATION OF SAMPLES FOR PHOTOMICROGRAPHY

The common laboratory chemicals found in high school chemistry stockrooms, photo stores, fertilizer distributors, and chemical supply houses provide an excellent source for samples. Most crystals are anisotropic and birefringent which means that they will refract plane polarized light emitted from the polarizer and will "bend" it until it is visible through the analyzer under cross-polarized illumination.

To prepare crystals for examination in the microscope, deposit a few milligrams of the appropriate chemical onto a glass microscope slide and carefully place a coverslip over the powder. Next, heat the bottom side of the microscope slide carefully with a Bunsen burner or hot plate until the powder has completely melted. Some chemicals decompose upon heating and will provide poor subjects for microscope examination. Salts are notoriously bad candidates for the melt-recrystallization process. The melting point of most salts is usually very high and any organic portion of the molecule usually undergoes degradation before melting. When melted, the molten chemical will flow underneath the coverslip and fill the entire volume between the coverslip and microscope slide. Either allow the slide to cool slowly before examination or place the melted chemical sandwich onto the microscope stage and examine the crystallization process occurring. Certain chemicals recrystallize very rapidly (within a few minutes) while others may recrystallize slowly over a period of days, weeks, months, and in some cases, years. Urea, sulfur, and benzoic acid are excellent examples of common laboratory chemicals that will recrystallize rapidly enough to be examined directly after melting. Some chemicals form more interesting crystals when cooled slowly over a period of hours or days. This can be accomplished by placing the microscope slide with it's melted chemical sandwich on a hot plate at 50°-70°C and reduce the temperature by 5°-10°C/hr over a period of several hours. By comparing the slowly recrystallized sample to one that was rapidly recrystallized, any differences in recrystallization can be noted. The interested photomicrographer should experiment with available chemicals to identify those that are optimal for microscopic analysis.

Another effective method of preparing crystals is to dissolve the chemical in a suitable solvent such as water, ethanol, acetone, or mineral spirits. A drop of the solution is then sandwiched between the microscope slide and coverslip and the solvent allowed to evaporate resulting in formation of crystalline patterns. This method is especially useful for chemicals in the salt family that usually decompose upon heating leaving a tar-like mess. In some instances, increasing the time for solvent evaporation will result in a dramatic increase in crystallite size.

The evaporation time can be controlled by partially sealing the coverslip with a bead of polymethylmethacrylate mounting medium (obtained from microscope dealers) leaving only a small edge of the coverslip exposed to the atmosphere. This will retard solvent evaporation which will allow a slower recrystallization process to occur usually resulting in larger, more well-formed crystallites.

Many of the colorful crystalline patterns illustrated later in this book derive from concentrated liquid crystalline DNA samples in which the aqueous solvent was allowed to evaporate slowly as described above. Under these circumstances a dilute solution of DNA will undergo a series of liquid crystalline phase transitions until a concentration is reached that renders the resulting crystallites extremely colorful and attractive. DNA exists in nature at high concentrations, and it is possible that liquid crystalline DNA phases have an invaluable function in biological systems. Again, many chemicals can display a wide spectrum of polymorphic crystalline patterns depending on whether they are melt-recrystallized or recrystallized by solution evaporation.

Samples for reflected light microscopy usually require little preparation. Reflected light microscopy can be likened to topographical surface examination with a high-power magnifying glass, and almost anything can be examined in microscopic detail with this technique. For example, the fine details of surface structure can be revealed on leaves, coins, printed paper, insects, and an endless variety of other specimens.

Perhaps the most interesting subjects for reflected light examination are integrated circuits. These doped silicon "chips" are generally packaged by either molding into an epoxy resin case or cemented into a ceramic case. It is virtually impossible to recover an integrated circuit from an epoxy resin case. During the manufacturing process, the resin flows onto the surface of the chip and penetrates into the etched microstructure. When the epoxy cases are broken or cracked open, the silicon often fractures through the center of the integrated circuit and all surface detail is lost. The cement in a ceramic case can, however, be scored with a hacksaw and split with a fine chisel to reveal the internal chip. Most programmable read-only memory (PROM), microprocessors, math coprocessors, and some random access memory (RAM) integrated circuits are protected with ceramic cases. It should be possible to build a substantial collection of integrated circuits using discarded computer parts.

Examination of integrated circuits with reflected light can serve many purposes. For instance, details of a particular circuit structure such as register areas, data busses, memory storage, and logic units are readily apparent. Also, by observing differences in the architecture of various integrated circuits, students can begin to get a handle on the complex electronics involved in modern devices such as radio, television, and computers. Later in this book, I illustrate a Nippon Electric Company (NEC) copy of the famous Intel 8080 microprocessor introduced in 1974 as the first stand-alone microprocessor. The introduction of this revolutionary microprocessor was a key factor in the early growth of a fledgling personal computer industry. The presence of different hues and shades of color, provided by rotating the first order red compensator plate when the microscope is in the reflected DIC mode, allows the eye to register more details of the integrated circuit than would be possible with ordinary reflected brightfield illumination. This consequence can be equated to the use of color graphics monitors on computers where color-coordinated screens have a more pleasing and restful effect on viewers than do monochrome screens with varying grey scales.

The biological stereo microscope can be useful in the examination of surface detail on integrated circuits. This is especially true of the larger microprocessors that may exceed $1 cm^2$ in area. By adding color filtration to the light sources for reflected light microscopy of integrated circuits, one may acquire colored highlighting effects that enhance the appearance of integrated circuits in photomicrographs. It is interesting to experiment with several light sources each with a different color filter. Quite a number of special highlighting effects can be obtained in this manner. Reflected light microscopy has also become an indispensable tool for the semiconductor industry due to its usefulness in characterizing manufacturing defects and monitoring the successive stages of integrated circuit fabrication.

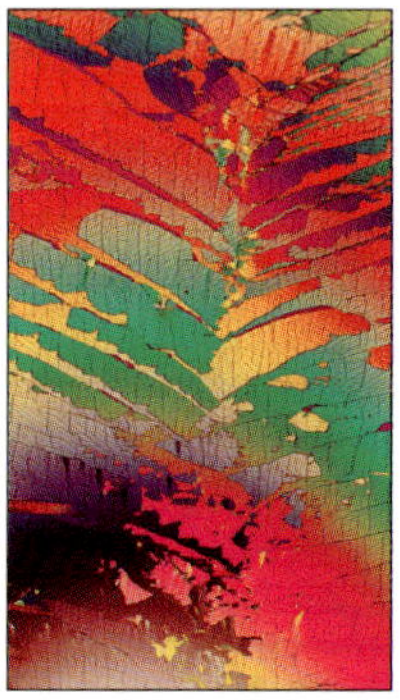

CREATIVE PHOTOMICROGRAPHY

By employing multiple exposure photomicrography, I have succeeded in generating a series of unusual micrographs which I have termed *microscapes*. Microscapes consist of multiple exposures (usually from 2 to 9) fashioned on a single frame of 35mm transparency film. These photomicrographs are intended to resemble novel, surrealistic alien-like landscapes and are designed to have the highest contrast and color saturation currently available with commercial photographic materials and processing techniques.

The microscape collection utilizes a variety of both classical and non-classical microscopy techniques. Each micrograph was fabricated using primarily the lower power 4x and 10x objectives (see Figure 3) where depth-of-field is maximized and the successive exposures remain in focus throughout the viewfield when thin crystalline samples are used. In certain instances, where the illusion of a great distance is intended, a thicker crystal sample or a higher (20x or 40x) power objective with decreased depth-of-field is employed. The crystals in the immediate foreground are brought into focus leaving a blurred (out of focus) background to simulate the actual effects generated by depth-of-field limitations with conventional photography equipment.

Selective masking of previously exposed areas is critical, in multiple exposure photography, to avoid undesirable overlap and washing out of successive exposures. Masks can be cut from a portion of black poster board and are placed over a selected area of the field lens (the lens allowing light to enter the microscope).

The first step in microscape construction is the exposure of a *foreground* which consists of a selected recrystallized chemical usually imaged with polarized light using the lower power objective and positioned so that only the bottom 30-50% of the film is exposed. Color transparency film is black when unexposed so the top half of the film remains black (unexposed) due to the crossed polarizers. A mask is then cut (from a portion of black poster board) to follow the profile of the first exposure and placed directly on the field lens (at the base of the microscope) to selectively cover and prevent the first exposure from receiving any additional light.

The second exposure can be an additional overlap on the first exposure, a crystalline formation resembling *mountains*, a *seascape* created with filters, or simply a simulated *sky*. Mountains can be simulated using a variety of recrystallized chemicals, although certain formations are more realistic than others. For instance, a snow-covered mountain effect can be obtained from a concentrated liquid crystalline solution of the polysaccharide xanthin gum in water. Other useful mountain backgrounds can be produced using crystals made from the organic buffer HEPES, from aspirin or sulfur crystallites, and from ampicillin dissolved then recrystallized from water. In some instances the foreground and mountain can be composed in a single crystal formation using recrystallized sulfur.

There are a variety of methods that can be applied to form diverse sky effects. After carefully cutting a mask to cover previous exposures, the microscope can be placed in the brightfield mode and a blue filter inserted into the light path (Figure 3). A very low-intensity (10–20 cc blue) filter will work best because the color saturation of the sky can be controlled by reducing or increasing exposure times. Very short exposures yield a dark blue sky while long exposures create a much lighter sky. It is important here that the mask conforms to the topography of all previously exposed regions of the film. On exceedingly irregular boundaries like crystalline mountain formations, a very short exposure time will result in a deep blue sky and avoid overlap of blue regions in the white mountains. A striking alternative to the blue filter technique involves cutting a 1 x 2 cm portion of a polyethylene sandwich baggy and stretching it longitudinally to approximately 1 x 3 cm. This action tends to align the chain-like polyethylene molecules, which enhances their birefringence. When viewed through cross polarizers, the aligned molecules combined with the thickness gradient created by the stretching, induces the sheet of polyethylene to generate a prism-like effect. When the field is defocused, the sharp lines diffuse resulting in a strikingly realistic yellow=>red=>blue *morning sky* effect. A third effect that I take advantage of is the stormy bluish-purple cloudy sky effect obtained by defocusing a bead of epoxy resin when imaged in polarized light with a 530 mm retardation plate positioned between the sample and the analyzer. Light diffracted by a mask accentuates the highlights producing an uncanny storm-like appearance. A blue filter can be added at the edge of the mask to make the effect more realistic.

At this point, a sun or moon can be added to the micrograph. This is accomplished by closing the field diaphragm of the microscope until the desired sun or moon diameter is reached. Next, the image of the shutter leaves is defocused until they merge to form a complete circle. After placing the appropriate filter (usually orange, yellow, or red) in the light path, the image of the field diaphragm can be relocated to any position in the view field simply by adjusting the centering thumbscrews on the substage condenser. Placing a mask over the bottom portion of the image will create a rising sun effect. With a yellow or orange filter, the exposure time length determines the outcome of color distribution.

Very short exposures produce images with a reddish perimeter and a saturated yellowish interior while longer exposure times tend to wash out the image to yield more of a moon-like effect. Long exposure times when using a red filter will cause a small yellowish center, while shorter times will result in an even color effect.

To simulate the reflection of the sun or moon on water, two methods can be applied. After the diaphragm image is correctly placed, a diffraction grating can be inserted into the light path to spread out the image and shift the image colors to longer (redder) wavelengths. Alternatively, a fine tooth comb inserted into the light path will produce a nice reflected moonlight effect. The new moon impression depicted in a microscape entitled "The Stand" is produced simply by inserting the tip of a ball-point in the light path. A yellow filter used here instead of orange produces a more realistic appearance.

Generation of stars and/or clouds is the final exposure in the series. After the field diaphragm exposure has been completed, a sealed microscope slide containing a solution of small liquid crystalline spherulites of the polypeptide poly-benzyl-l-glutamate is placed on the microscope stage. At low magnifications (10x), the spherulites appear as point-sources of light. By leaving the field diaphragm closed, an area on the slide can be located that is devoid of spherulites and this area is placed directly over the previously exposed diaphragm image area. This prevents stars from being imaged in the center of the moon or sun. After recentering and fully opening the field diaphragm, the exposure is made. Almost any crystallite preparation that produces small isolated crystals can be substituted for the polypeptide preparation. I have found that ascorbic acid (vitamin C) will produce suitable isolated crystallites when dilute solutions in ethanol are evaporated quickly. Many other preparations will probably produce acceptable star candidates, although the best results are usually obtained from evaporation of dilute solutions of the appropriate chemical. Clouds are produced by defocusing colorless birefringent crystals of Cibachrome bleach. These crystals are usually imaged in the top-most portion of the view field and exposed for long times to wash out all color.

Perhaps the most difficult microscapes to produce are the beach scenes where the morning sky and ocean are created in a single exposure. This is accomplished by placing a blue filter over the mask at the field lens and defocusing the microscope until a sharp boundary occurs at the junction of the blue filter and the stretched polyethylene image. Correct positioning of the field diaphragm image in order to create a rising sun effect, for example, can be especially difficult when one attempts to construct a certain beach scene.

When I observe an unusually promising crystalline formation I usually compose enough images to fill several 24-exposure magazines of film. In this case, I proceed through the sequence as outlined above for the first series of exposures (to produce the intended microscape), then reverse the procedure constructing another similar microscape thus minimizing the number of substage condenser centering operations. Because the sun or moon is not relocated during recording of the two successive microscapes, almost identical copies are obtained. This is especially beneficial if the resulting arrangement of exposures is optimal. With the proper equipment on hand for the assembly of microscapes, the interested photographer is limited only by the boundaries of his (or her) own imagination.

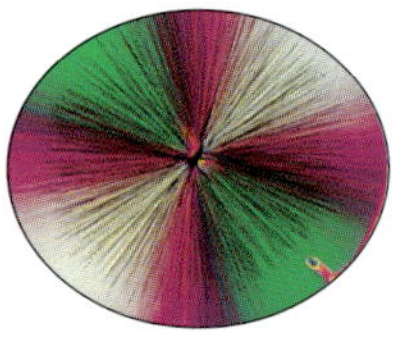

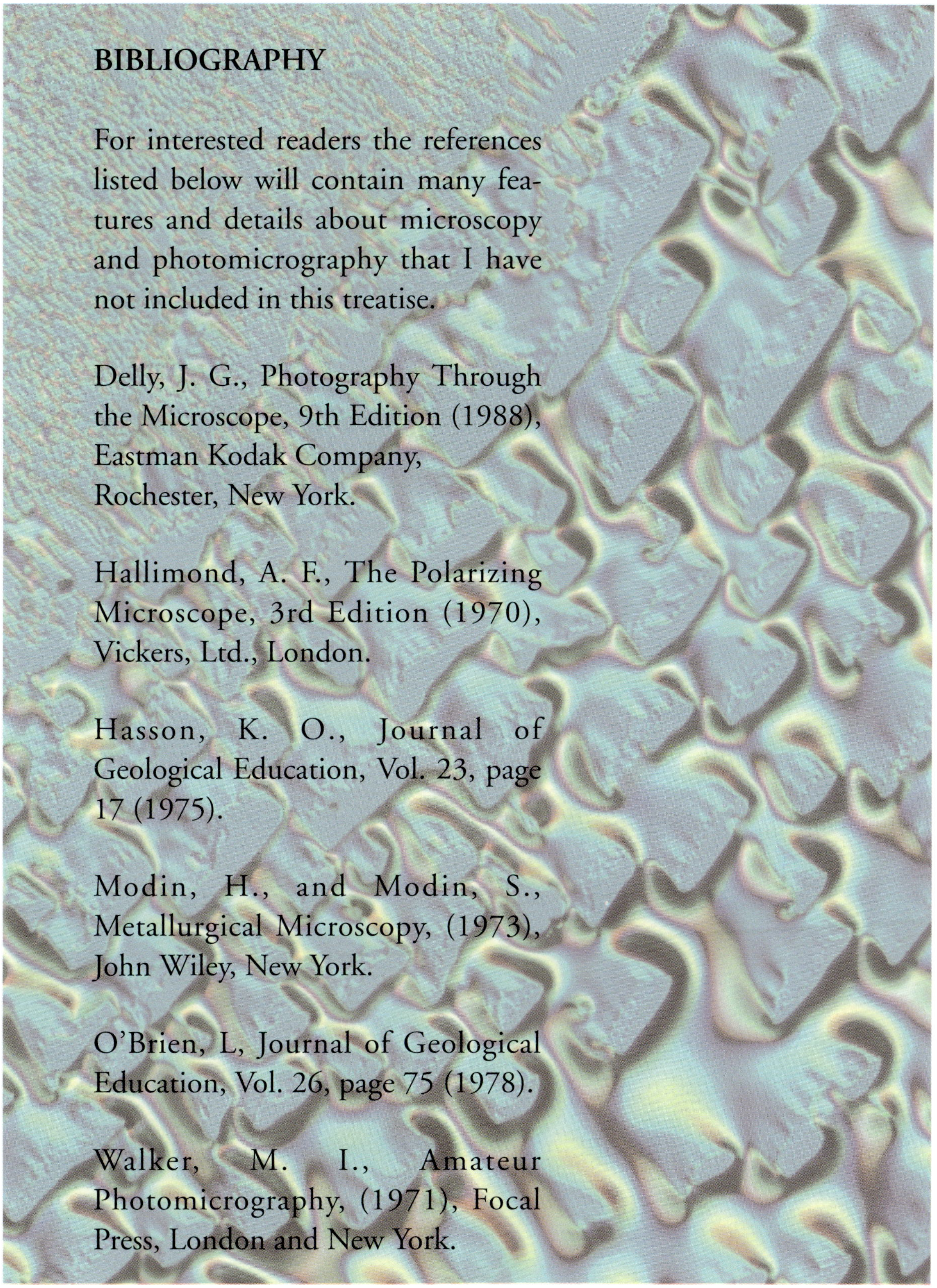

BIBLIOGRAPHY

For interested readers the references listed below will contain many features and details about microscopy and photomicrography that I have not included in this treatise.

Delly, J. G., Photography Through the Microscope, 9th Edition (1988), Eastman Kodak Company, Rochester, New York.

Hallimond, A. F., The Polarizing Microscope, 3rd Edition (1970), Vickers, Ltd., London.

Hasson, K. O., Journal of Geological Education, Vol. 23, page 17 (1975).

Modin, H., and Modin, S., Metallurgical Microscopy, (1973), John Wiley, New York.

O'Brien, L, Journal of Geological Education, Vol. 26, page 75 (1978).

Walker, M. I., Amateur Photomicrography, (1971), Focal Press, London and New York.

ACKNOWLEDGEMENTS

I like to extend my deepest appreciation to Professor Jack E. Crow, director of the National High Magnetic Field Laboratory in Tallahassee, Florida. Without the fiancial and moral support provided by Professor Crow, my photomicrography career would have died in it's infancy. I also like to thank my technician, Bonnie Hendrickson, who has continued to produce quality photographs under the most dire of circumstances. Other friends and co-workers at the Magnet Lab and Florida State University's Center for Materials Research and Technology have also provided needed friendship and assistance. These are: Walter Thorner, Janet Patten, Jim Ferner, Dena Stephenson, Kyle Orth, Dr. Joe Schlenoff, Dr. Randolph Rill, Dr. Michael Kasha, Dr. David Lind, Tom Fellers, Heather Allen, and Jack Kane. Last, but certainly not least, I like to thank my wife, Deborah, who has had the toughest job of all, tolerating me.

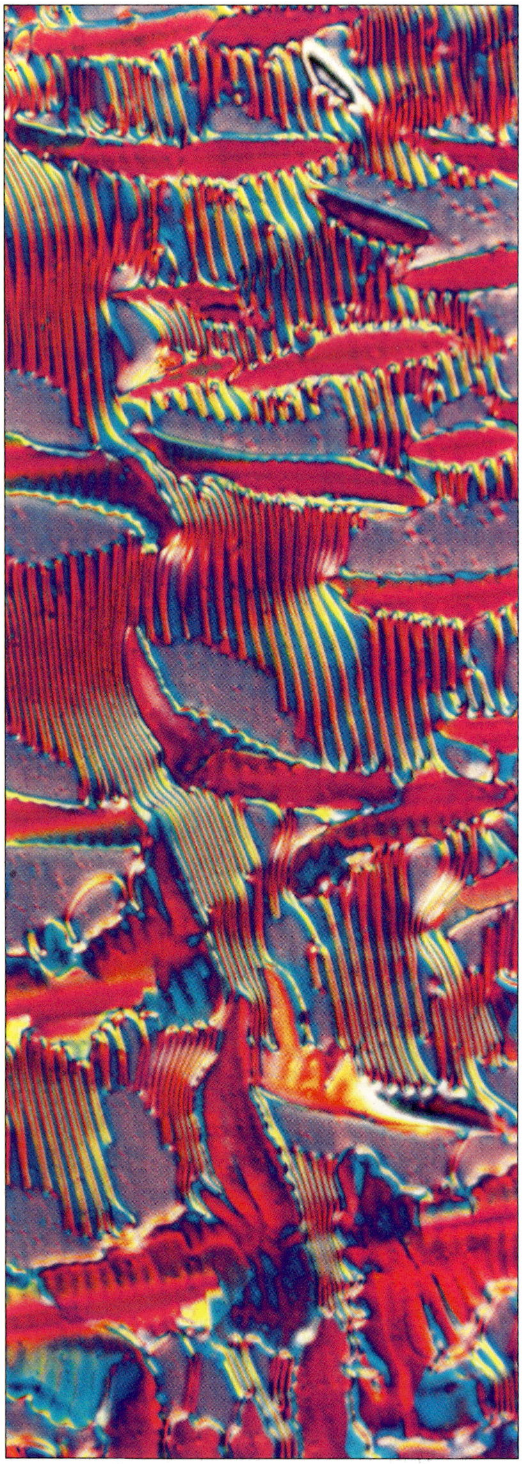

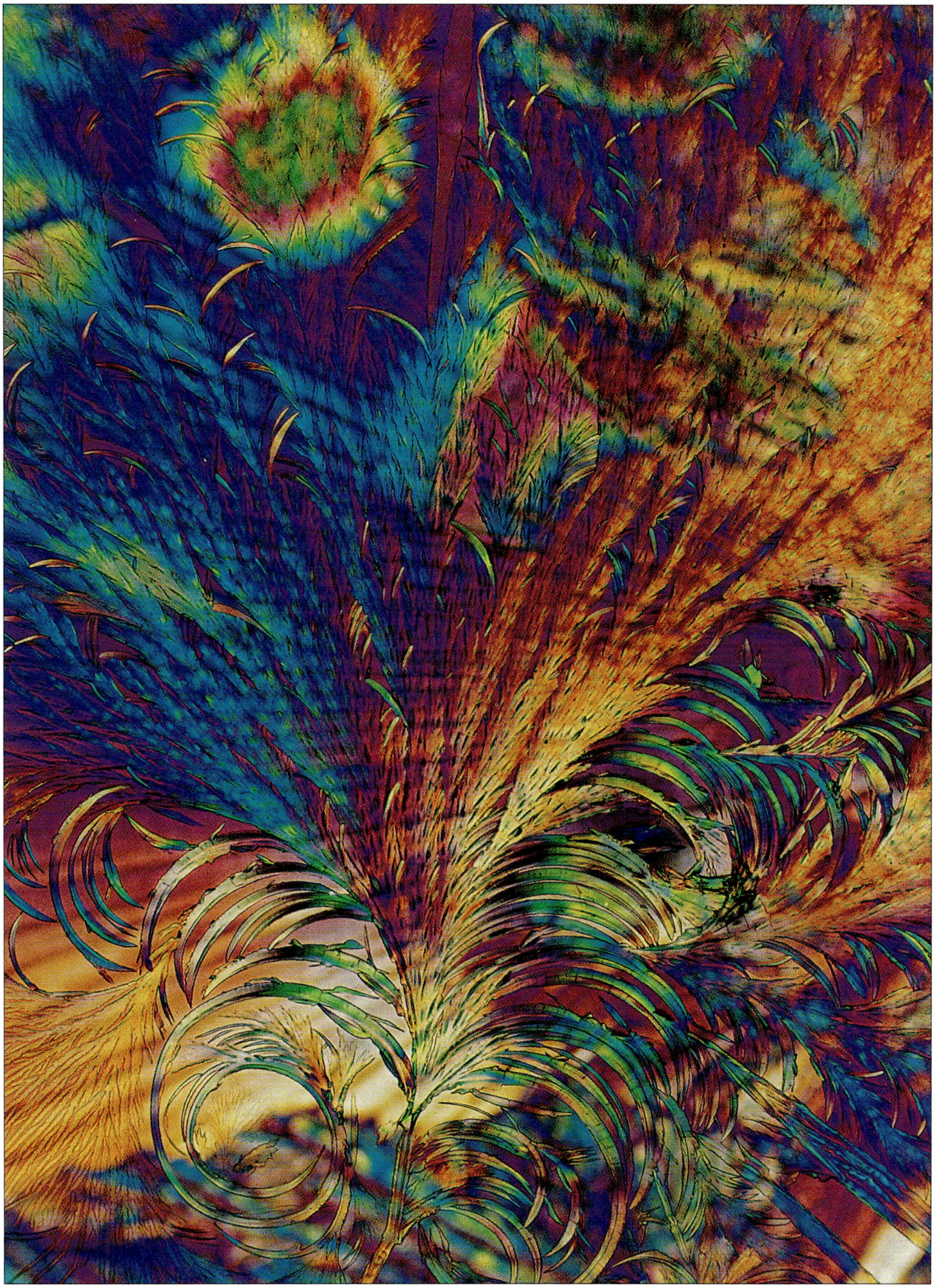

AZT

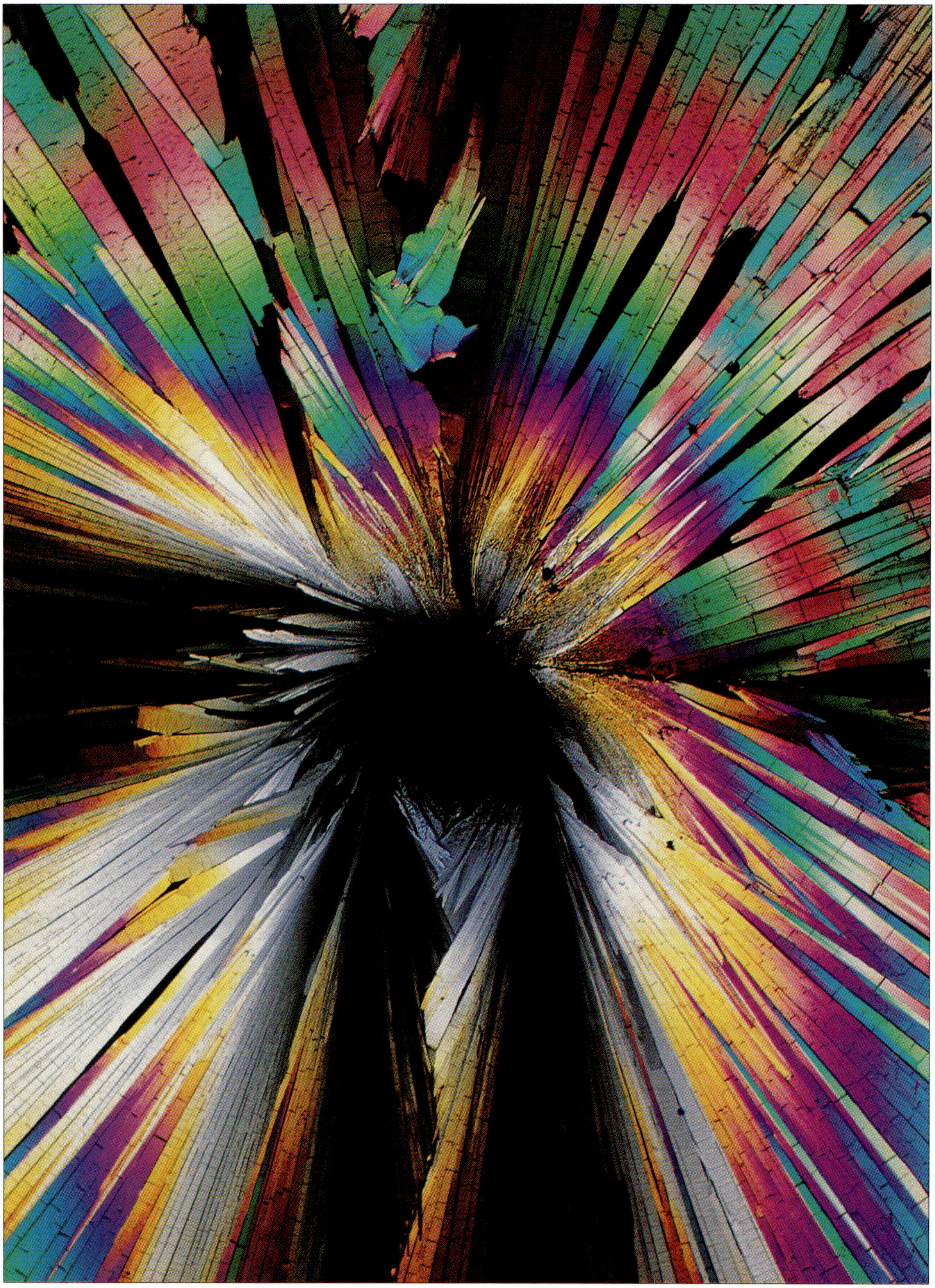

p-Aminobenzoic Acid (PABA)

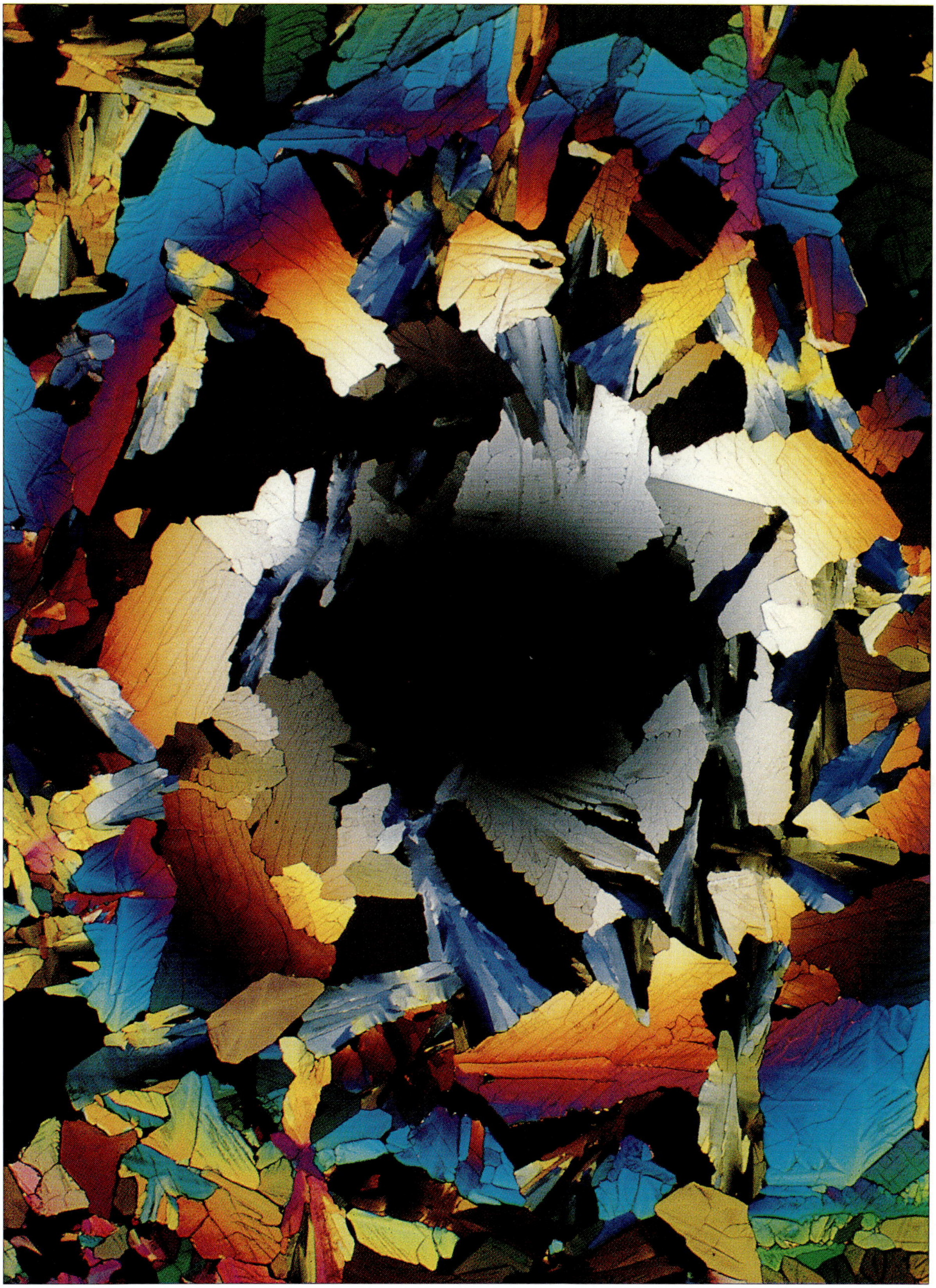
Beta Carotene

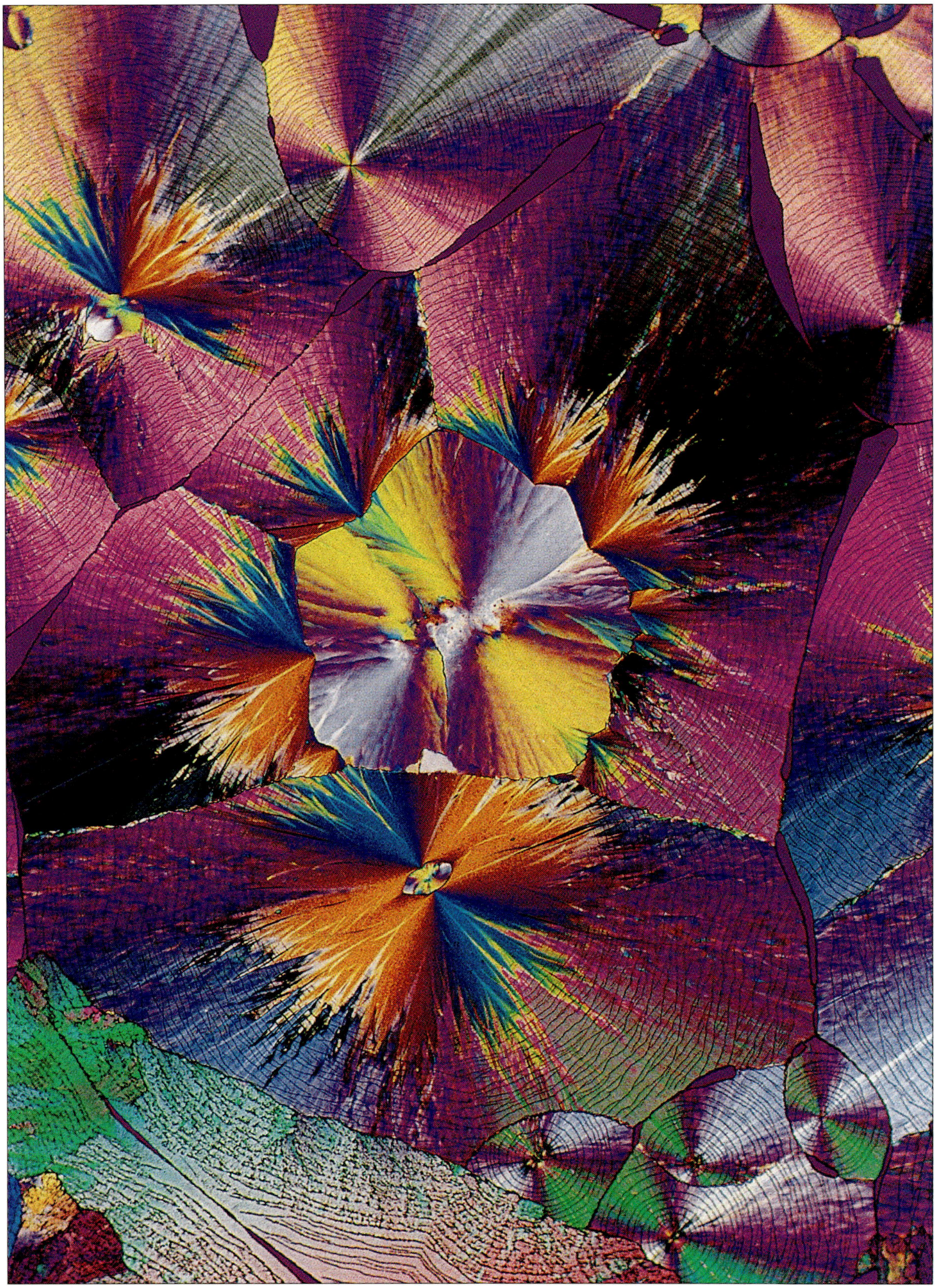

Niacin

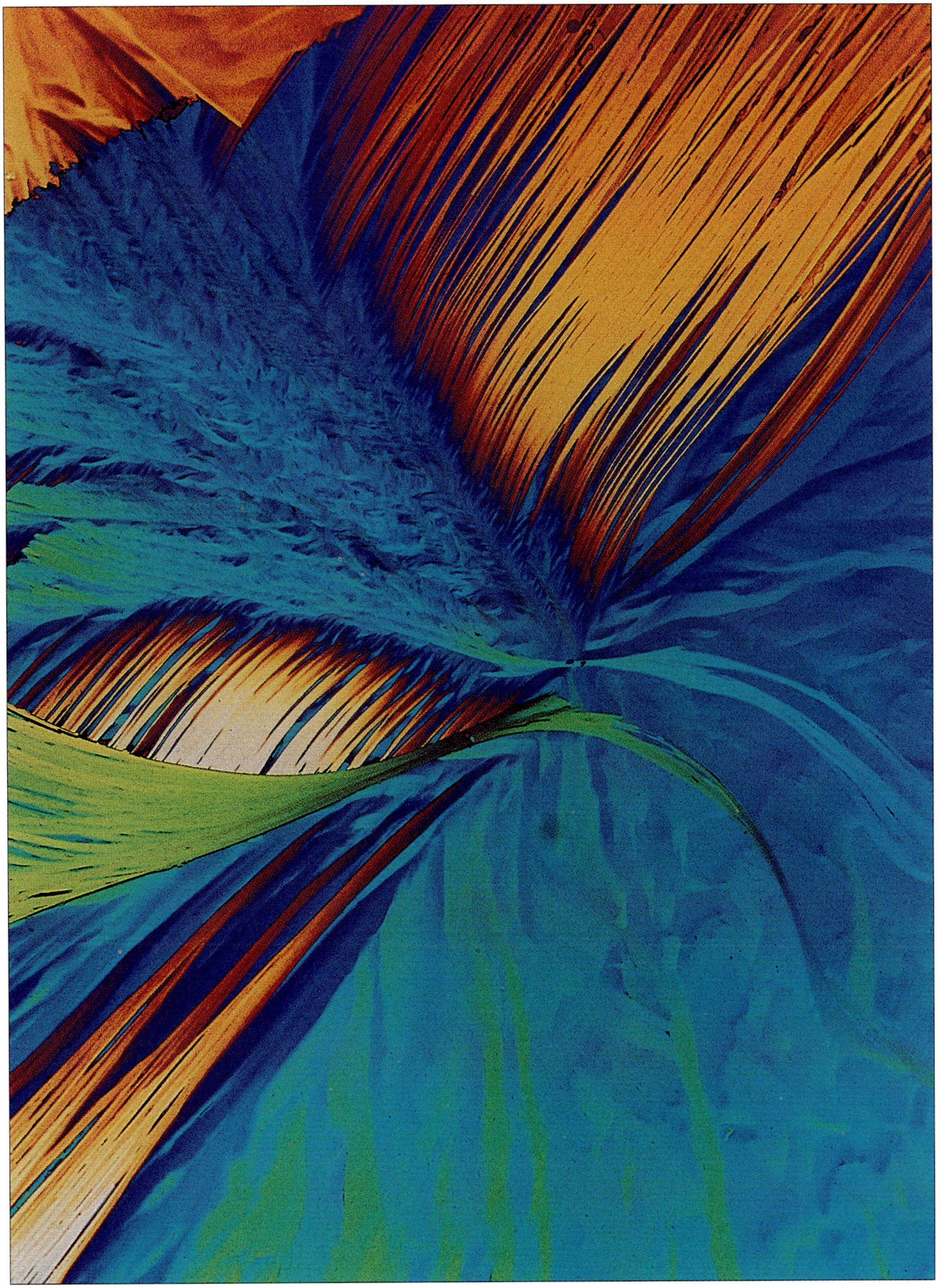
Anthranilic Acid

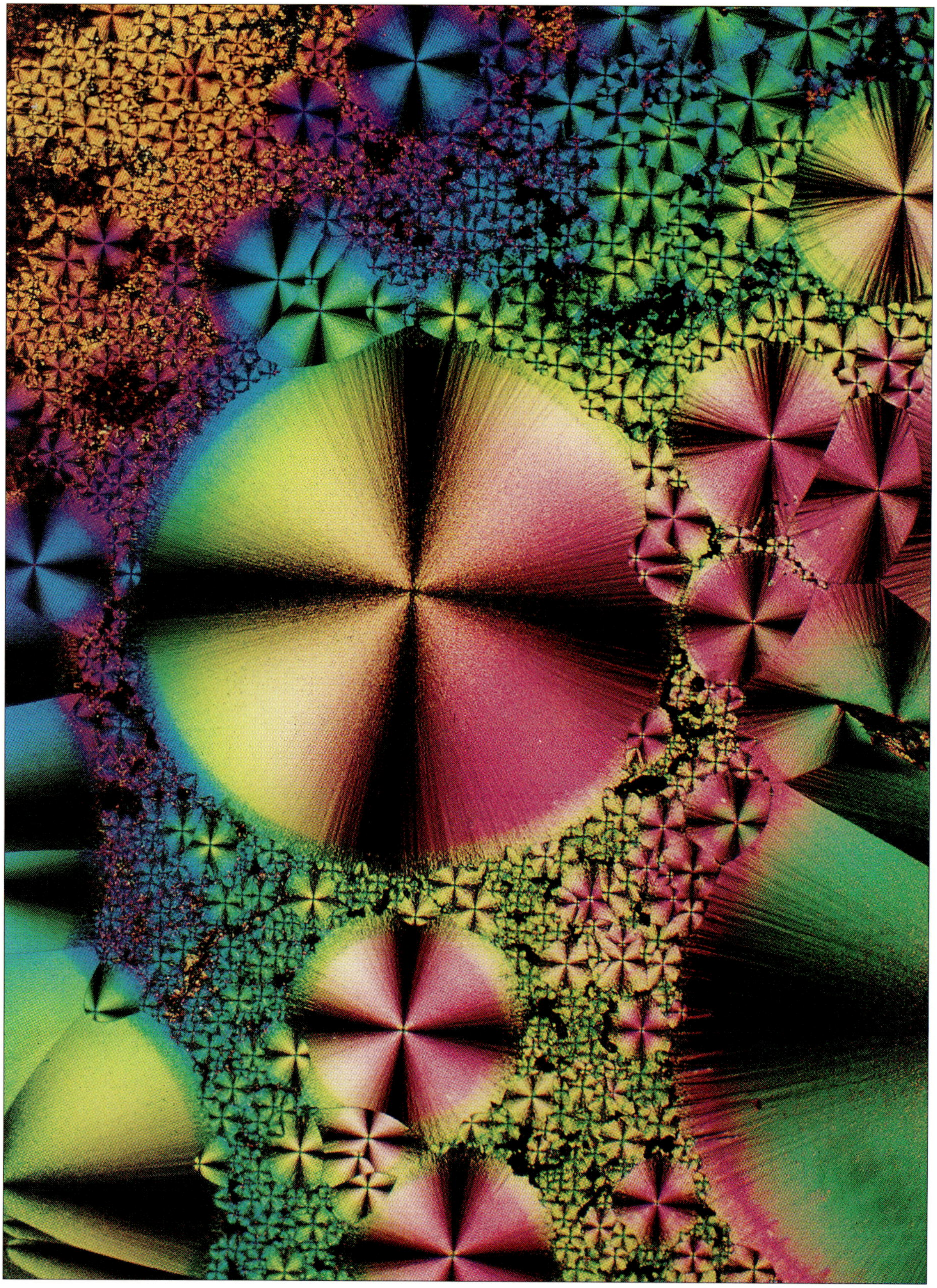

Thiamine (Vitamin B1)

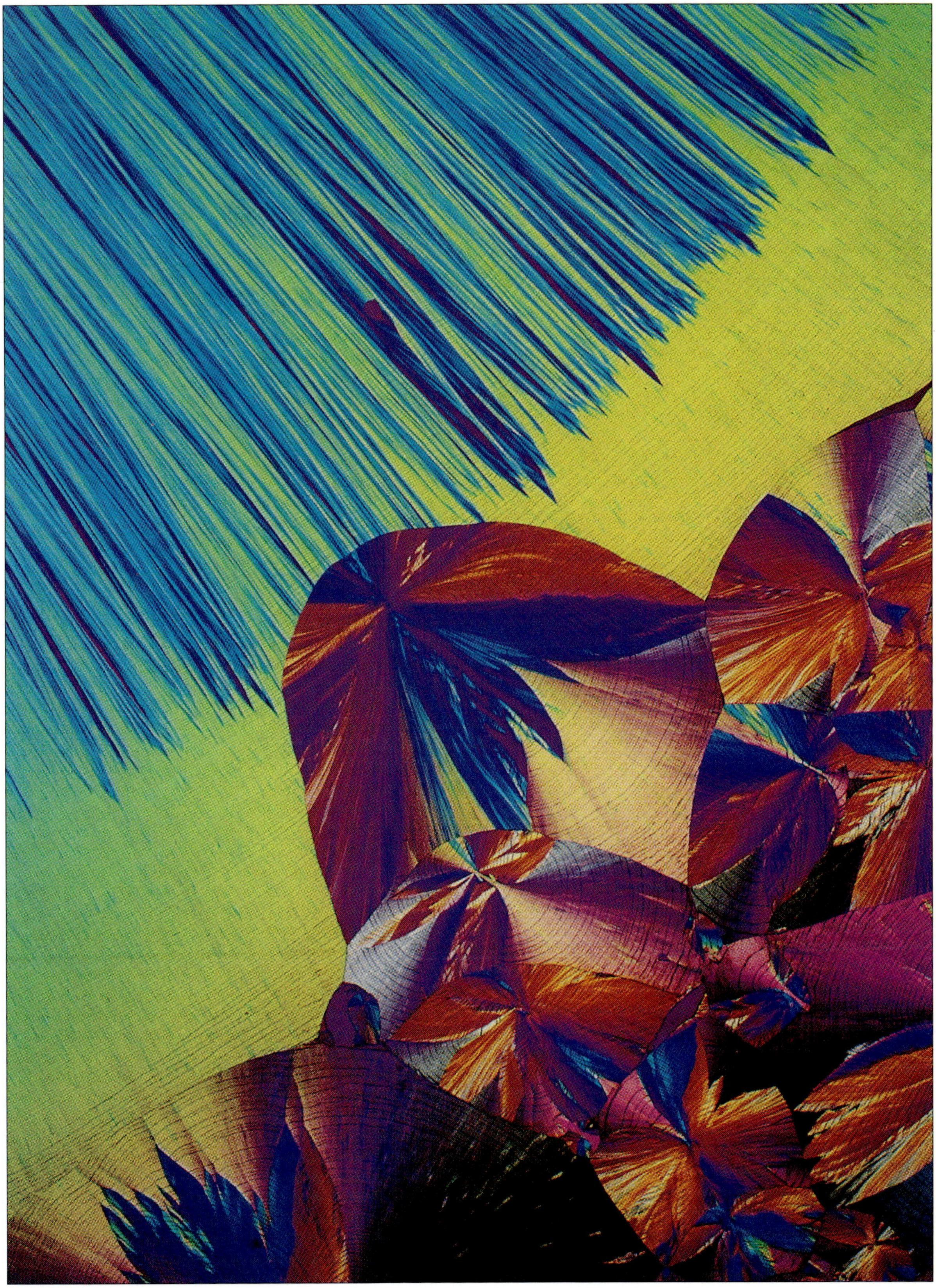

Retinoic Acid (Vitamin A)

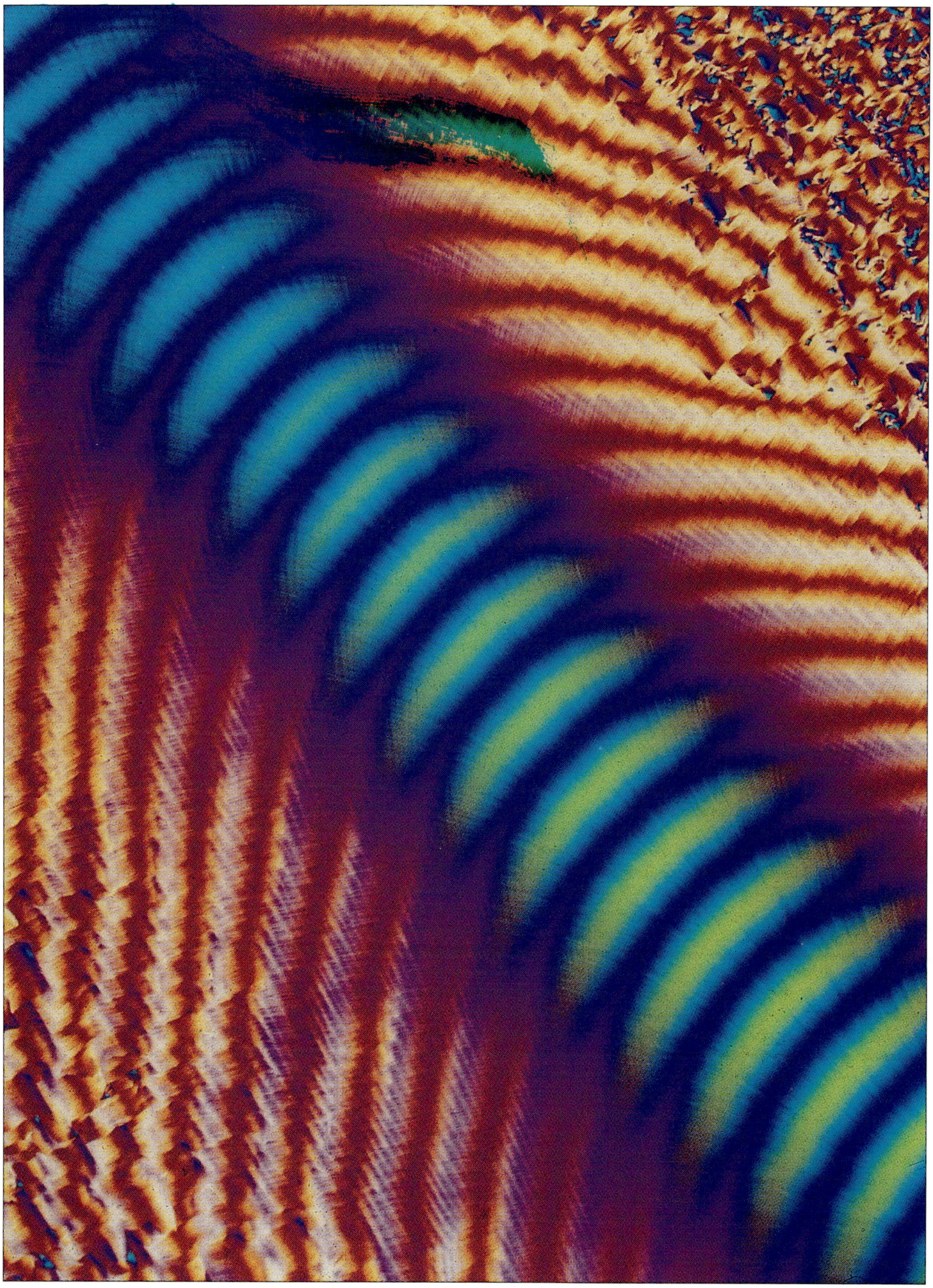

Riboflavin (Vitamin B2)

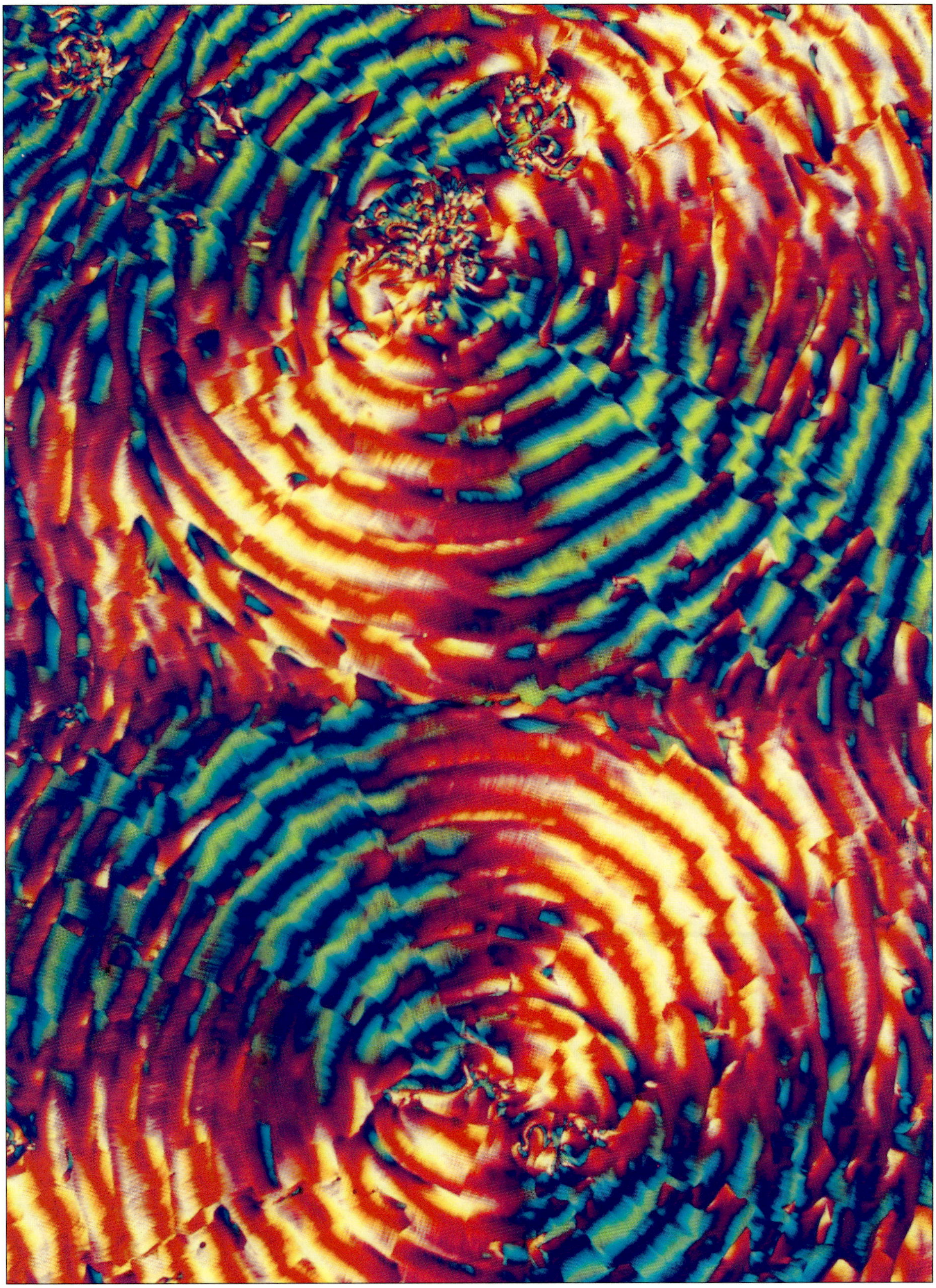

Riboflavin (Vitamin B2)

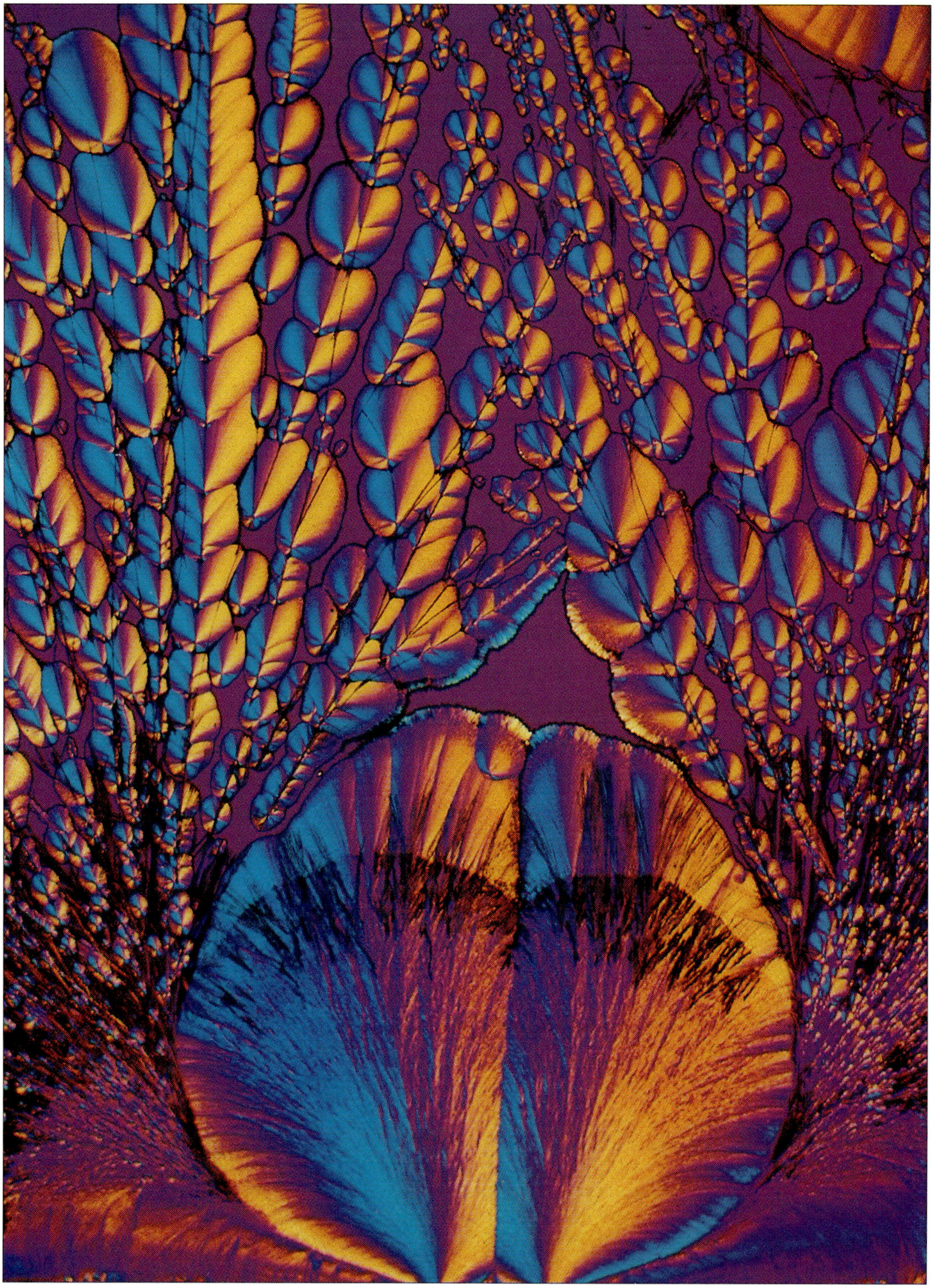

Pantothenic Acid (Vitamin B5)

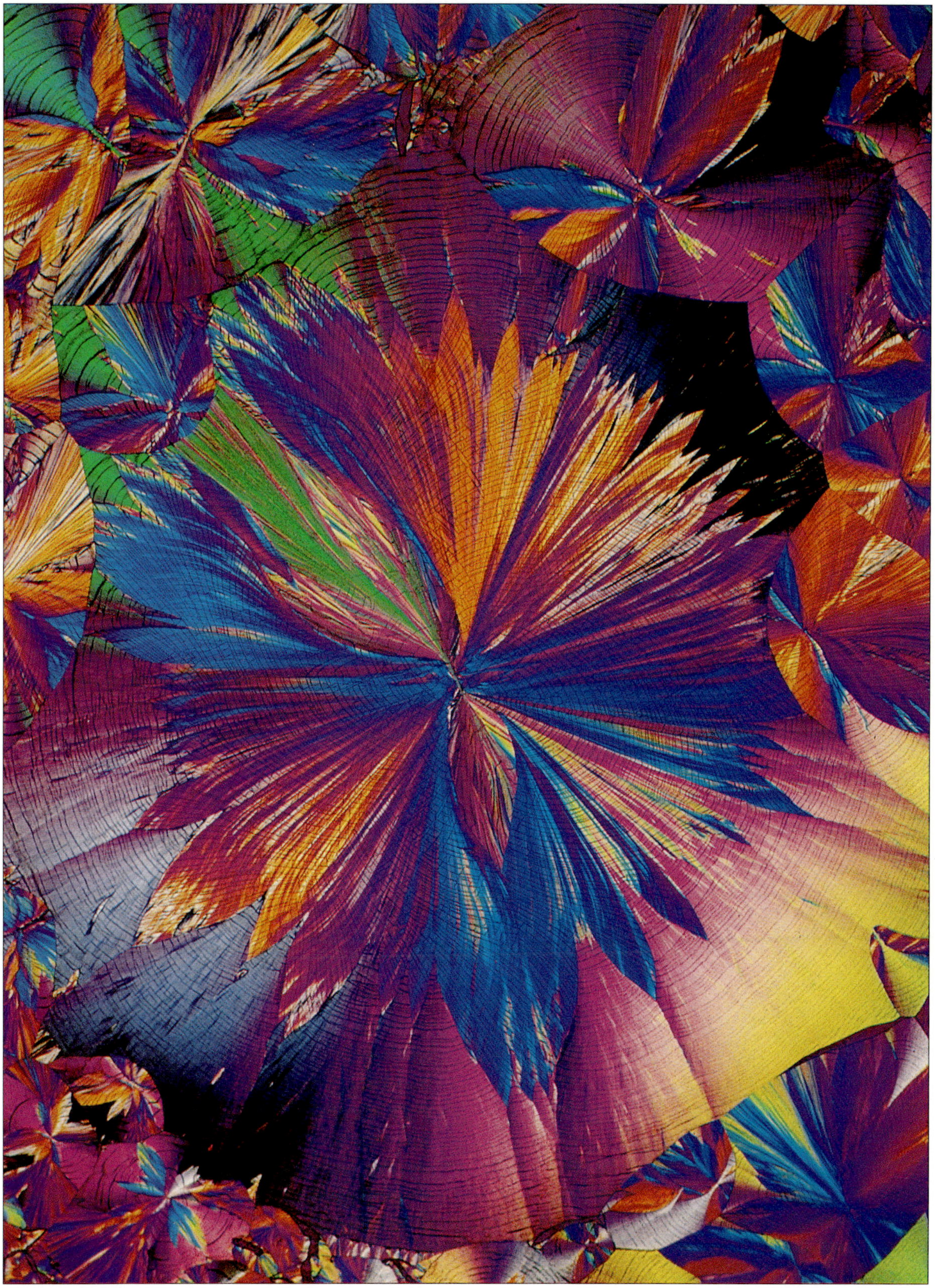

Retinoic Acid

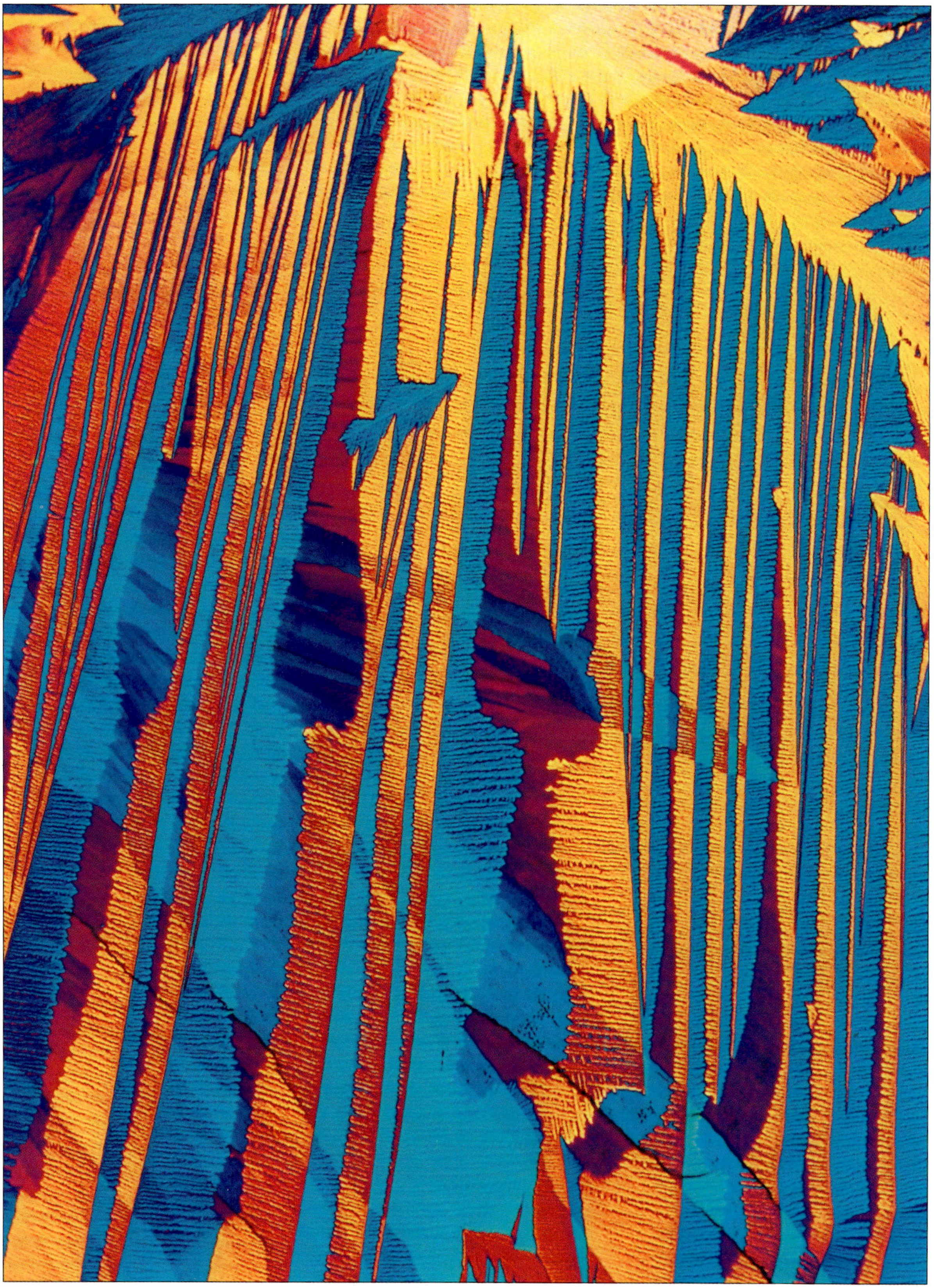
Folic Acid (Vitamin M)

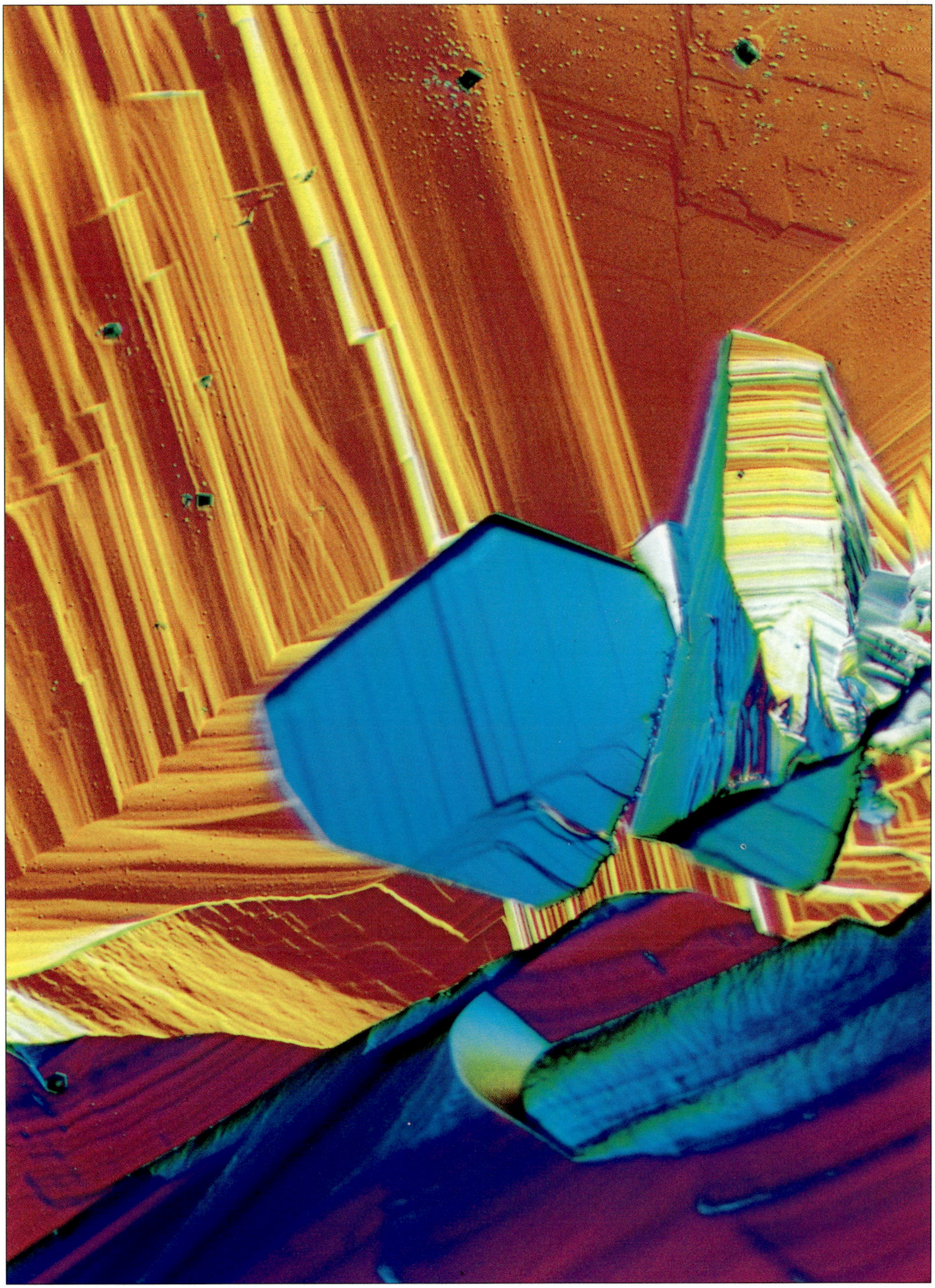

Single Crystal Surface

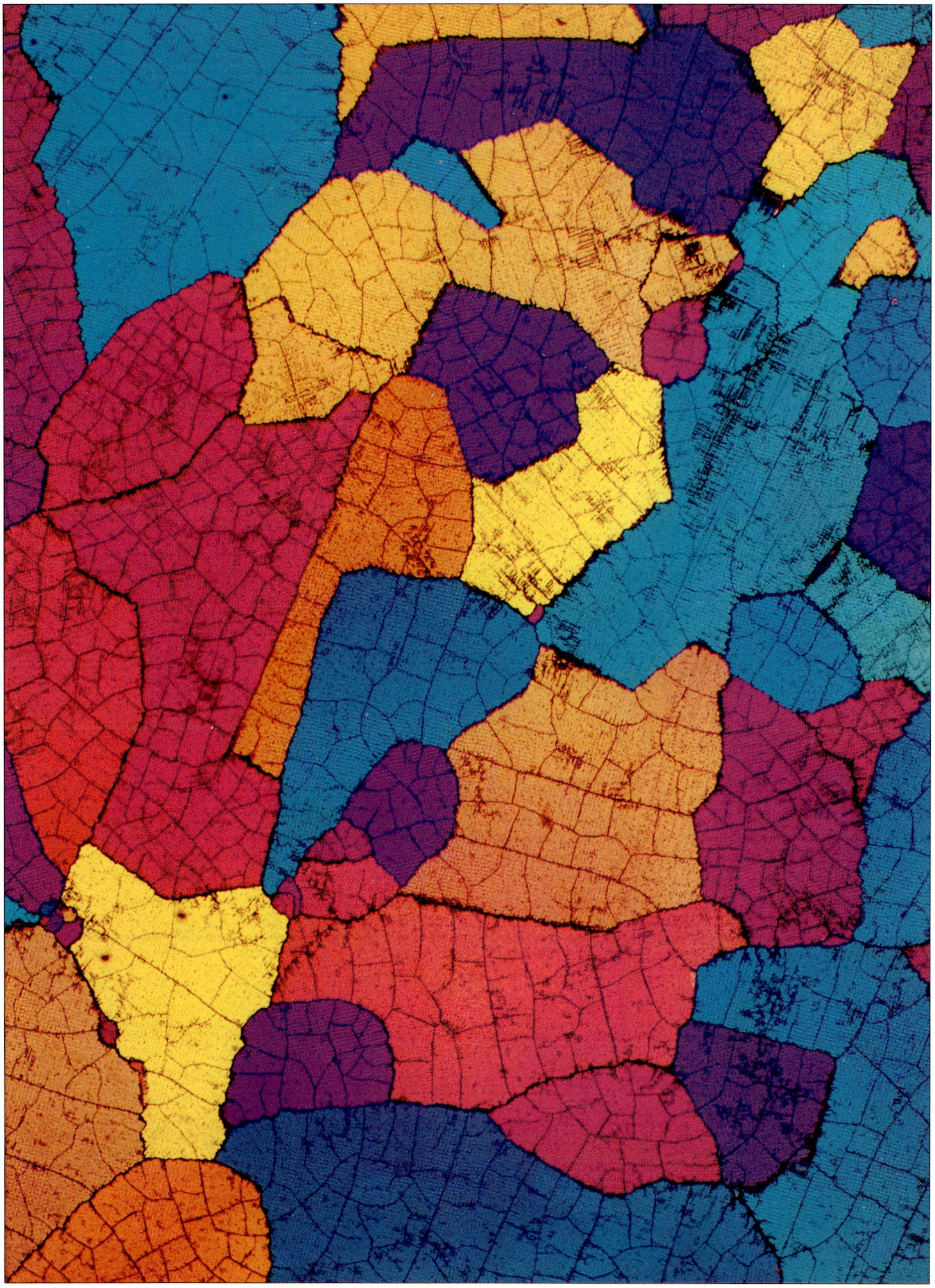

Calciferol (Vitamin D2)

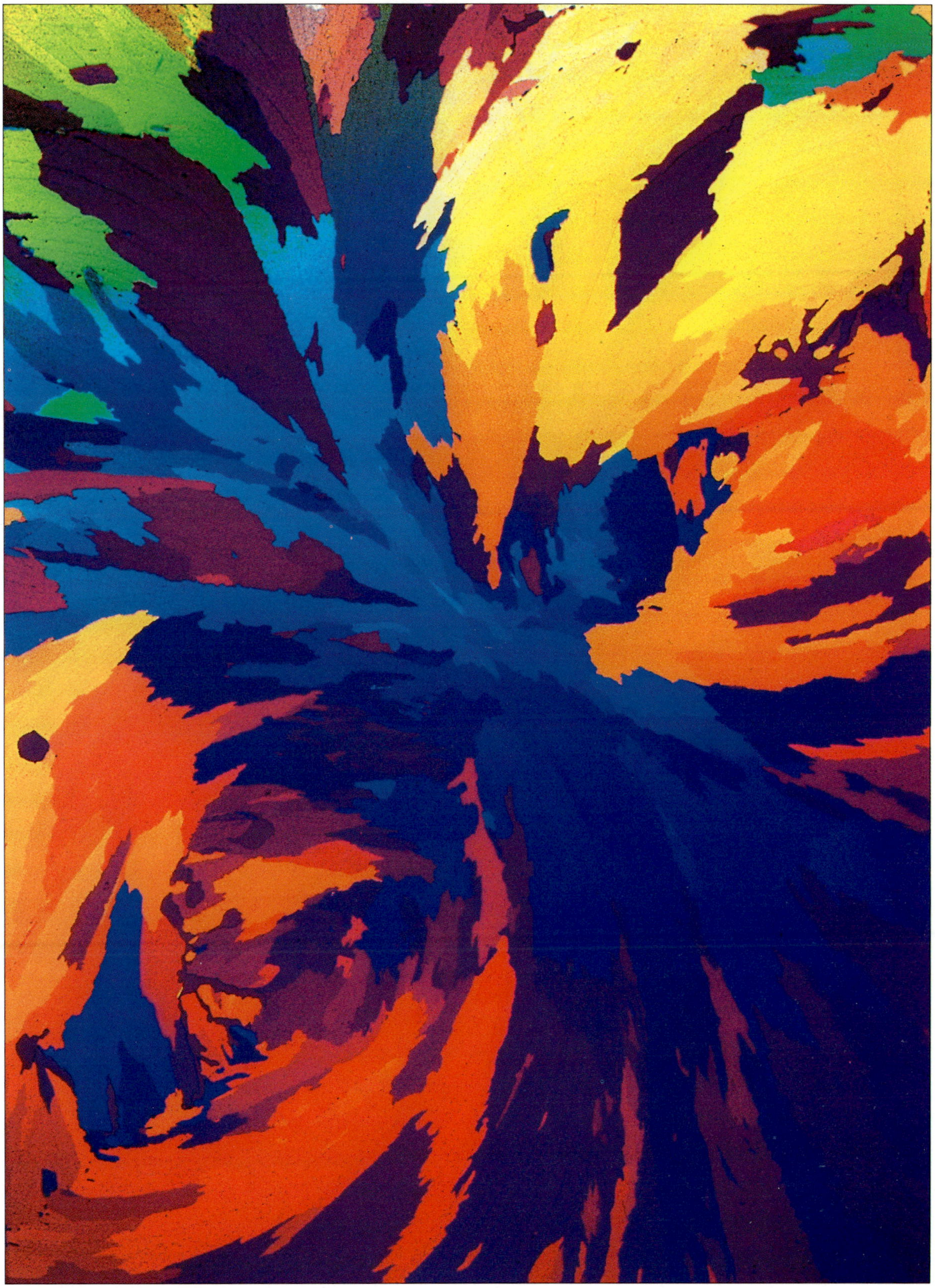

Thiamine (Vitamin B1)

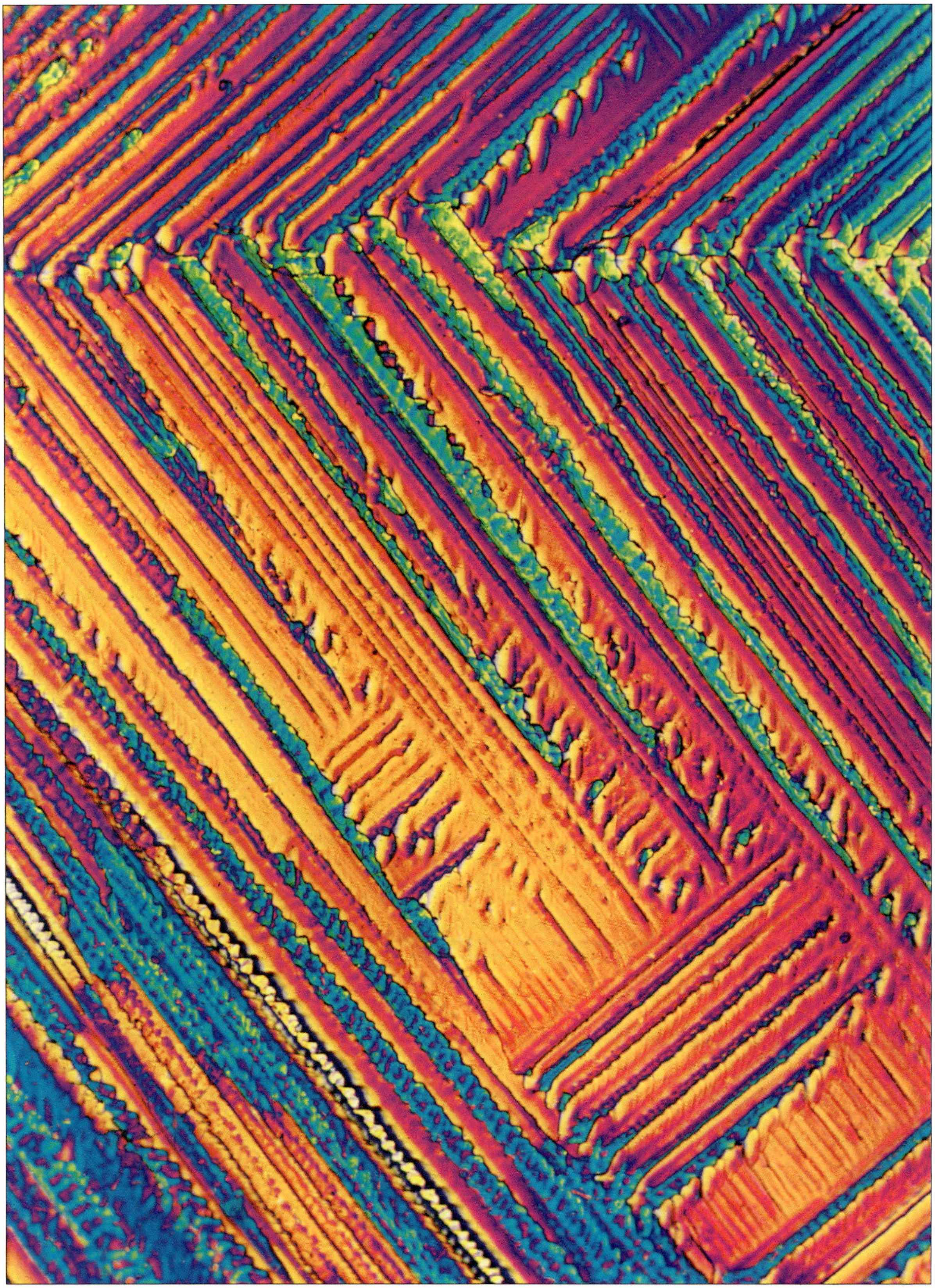

Adenine (Vitamin B4)

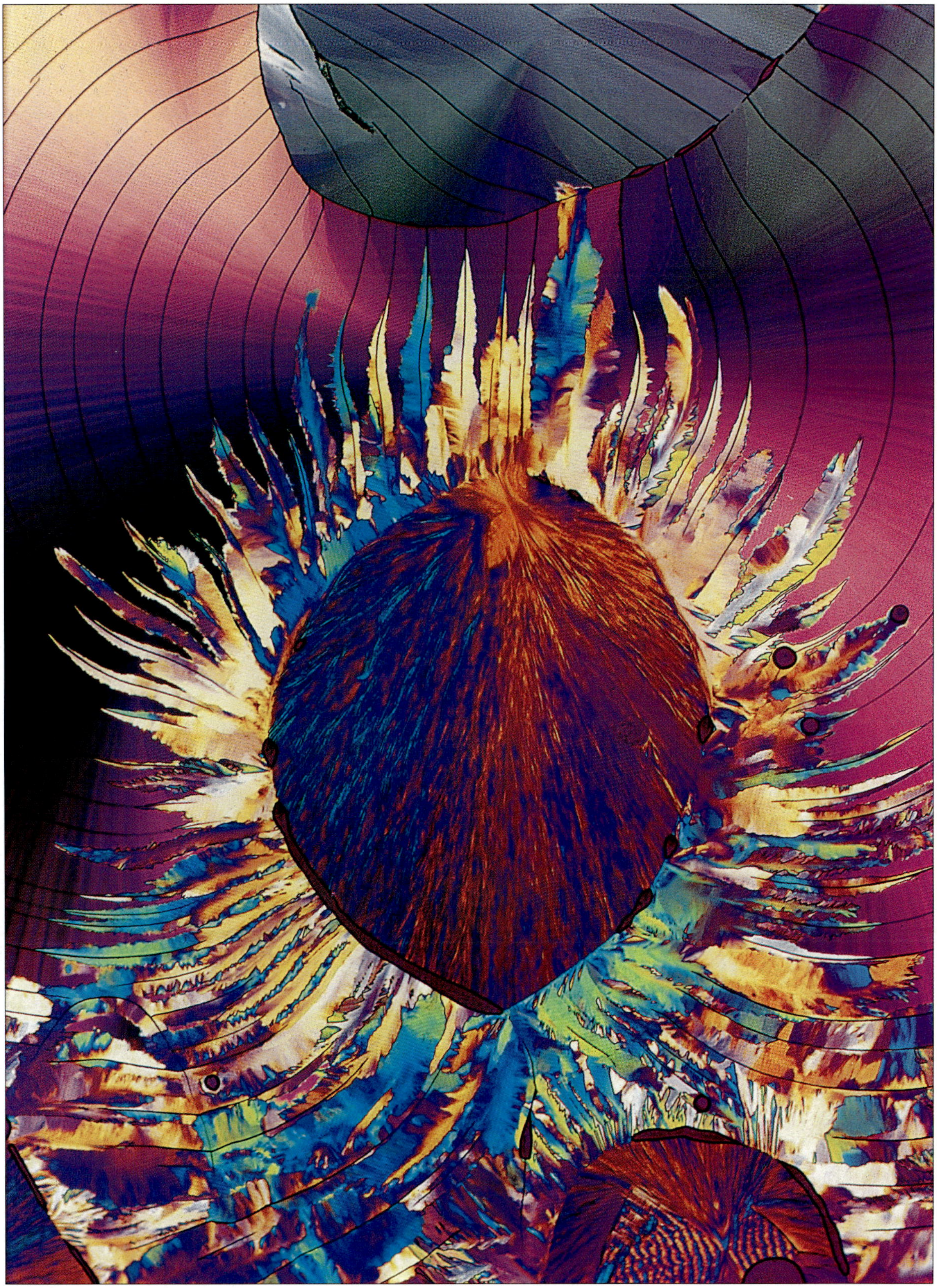

Retinoic Acid (Vitamin A)

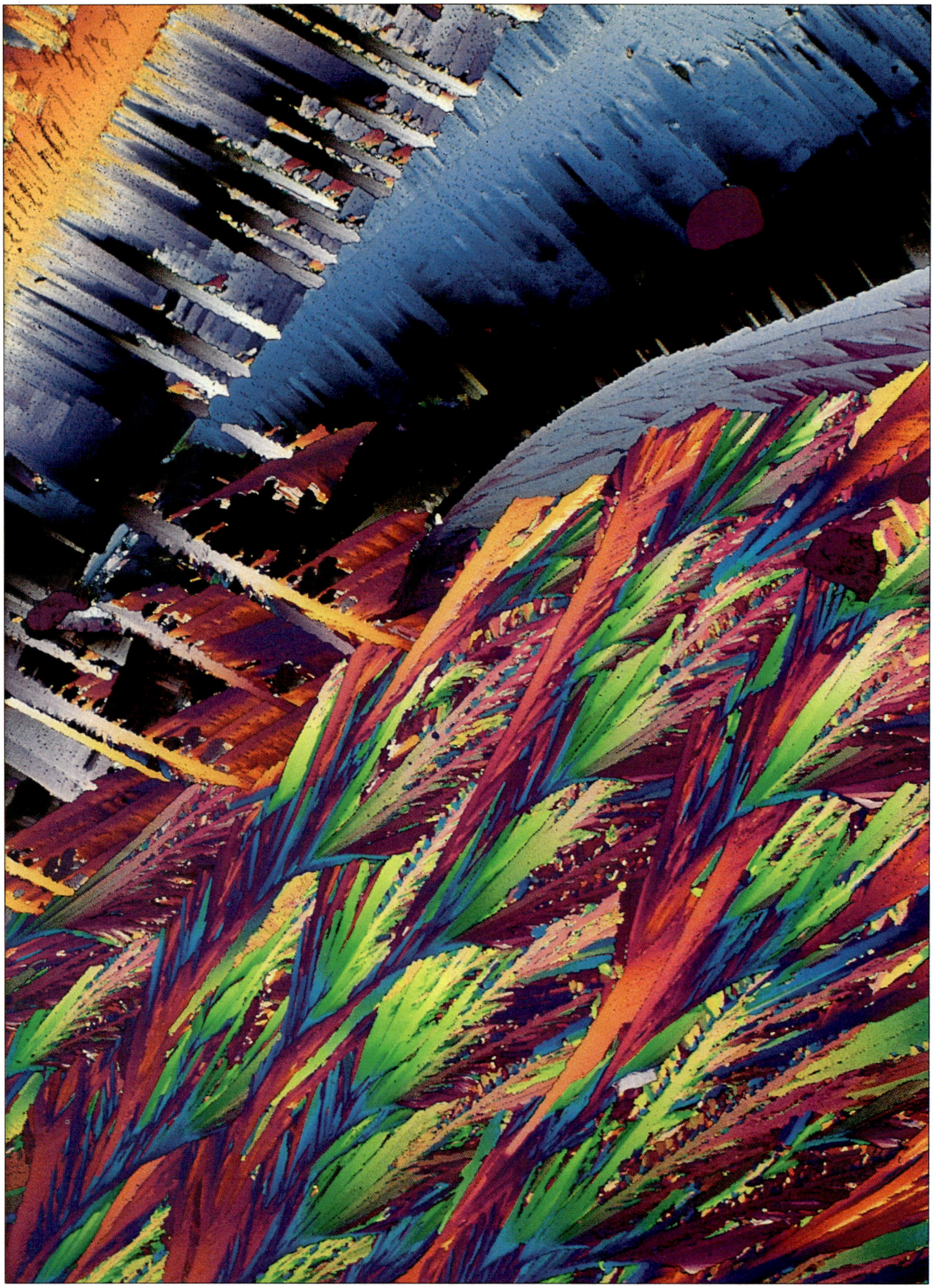

Beta-Carotene

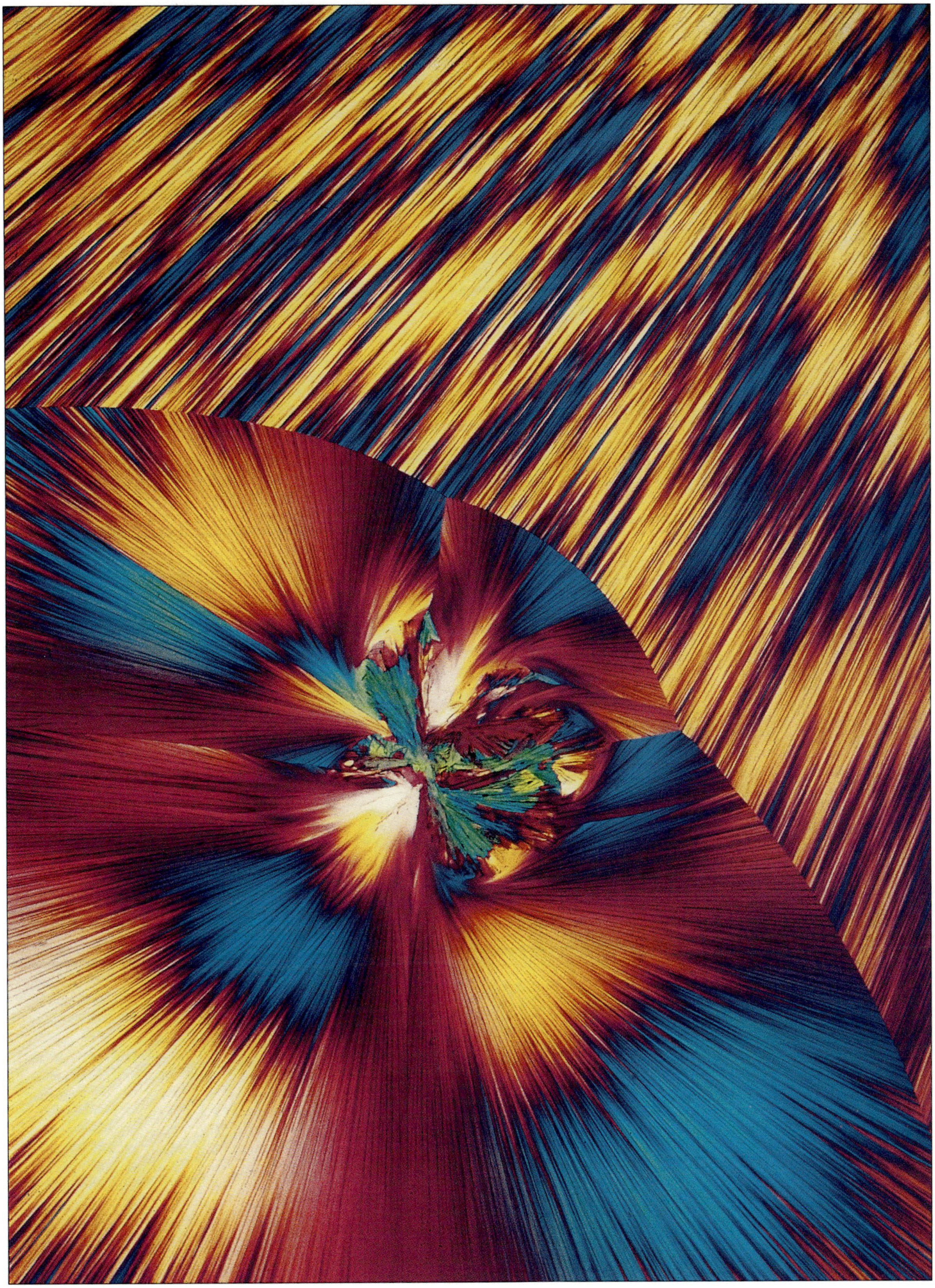

Cholesterol

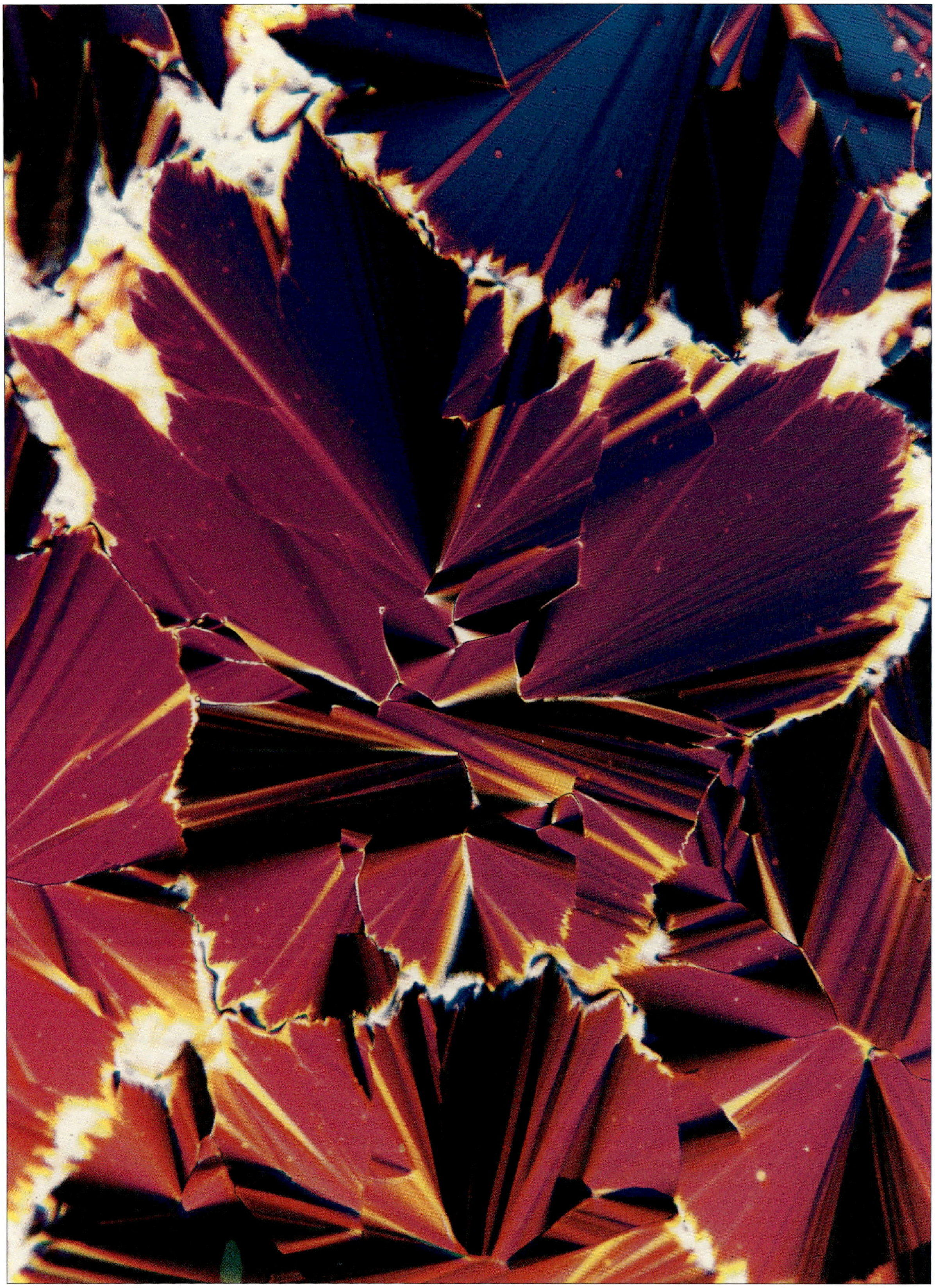

Liquid Crystalline DNA

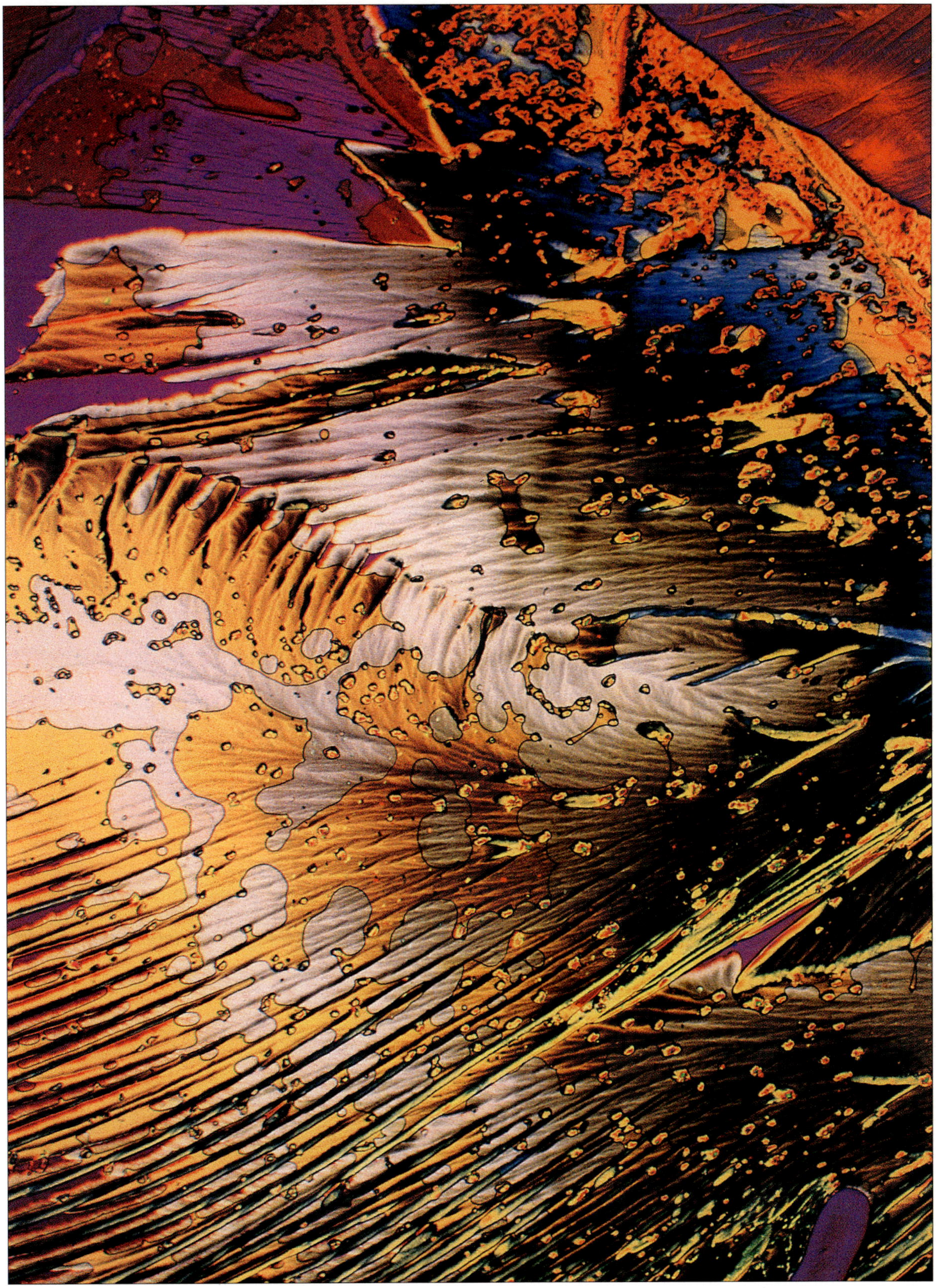

Ascorbic Acid (Vitamin C)

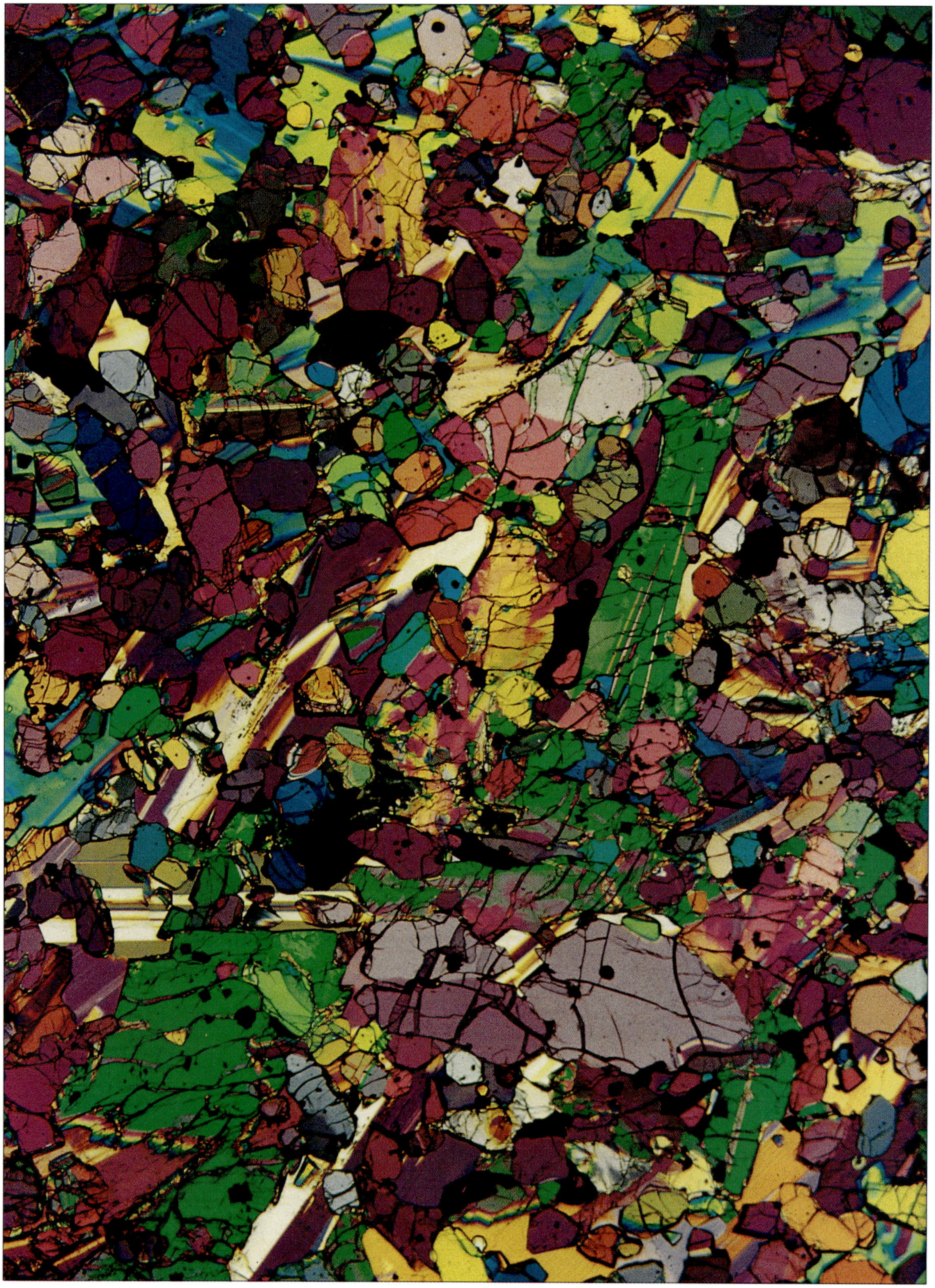

Apollo 12 Moon Rock

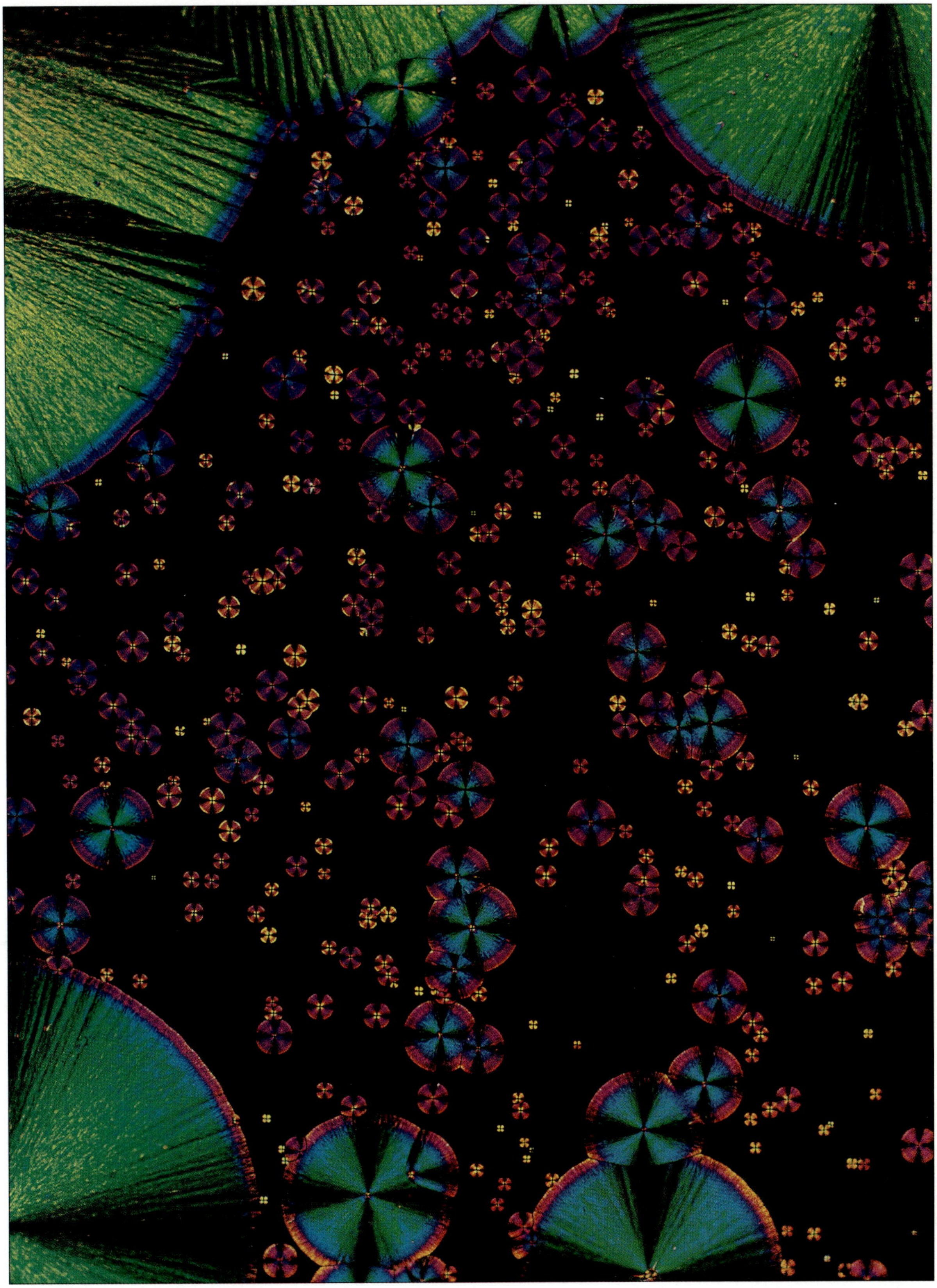

Thiamine (Vitamin B1)

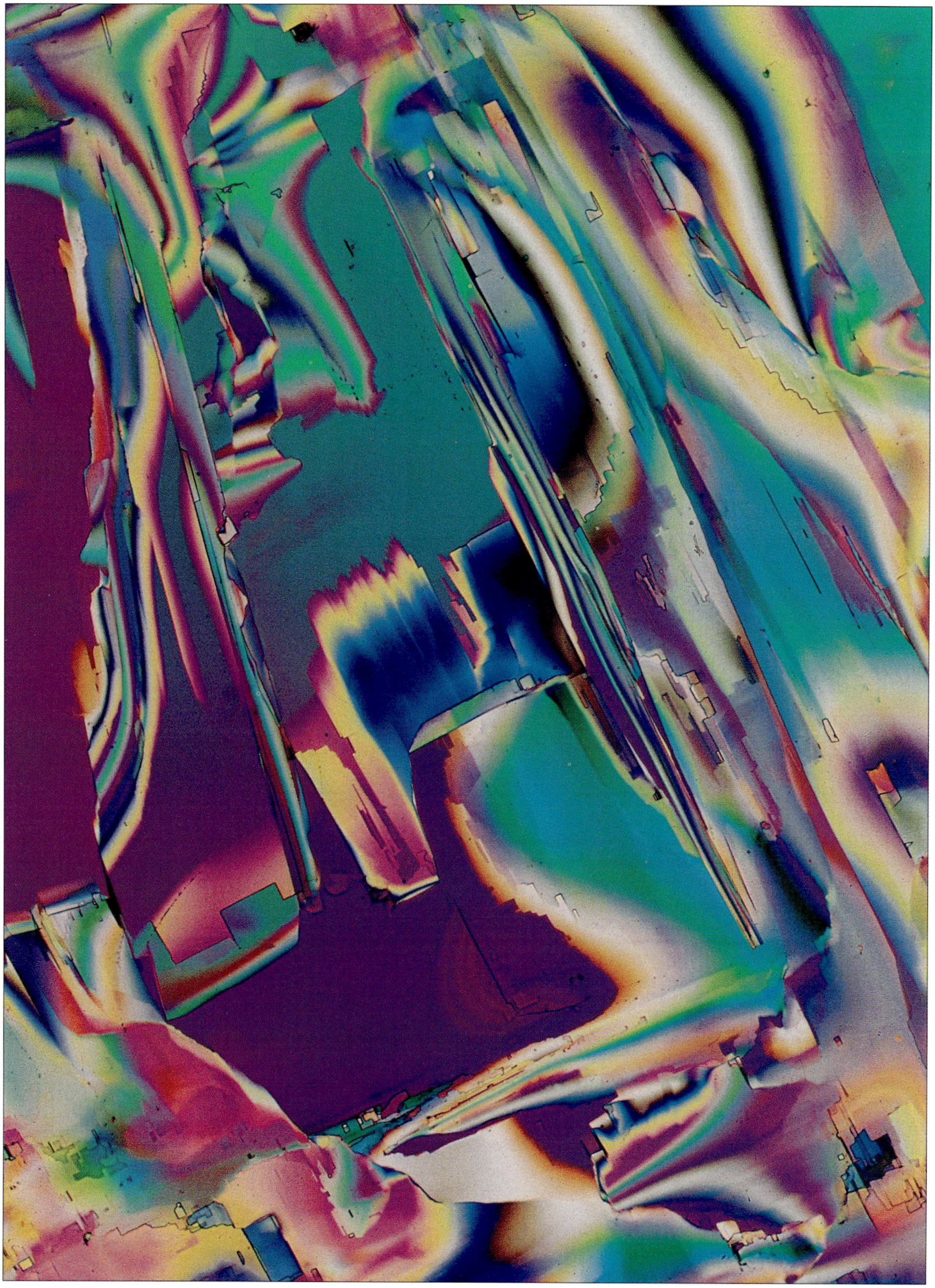

Tacrine (Anti-Alzheimers)

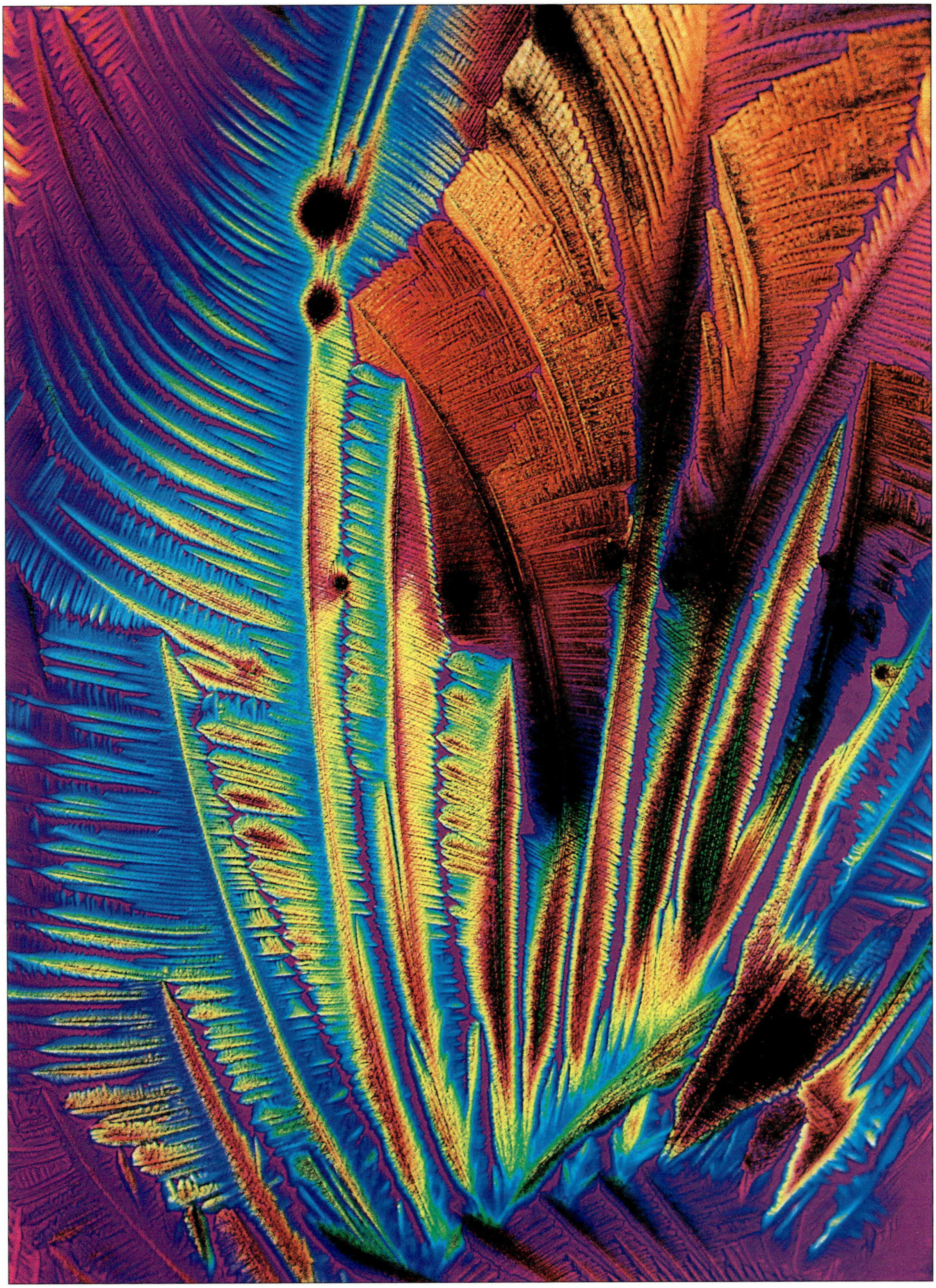

DDI (Anti-AIDS)

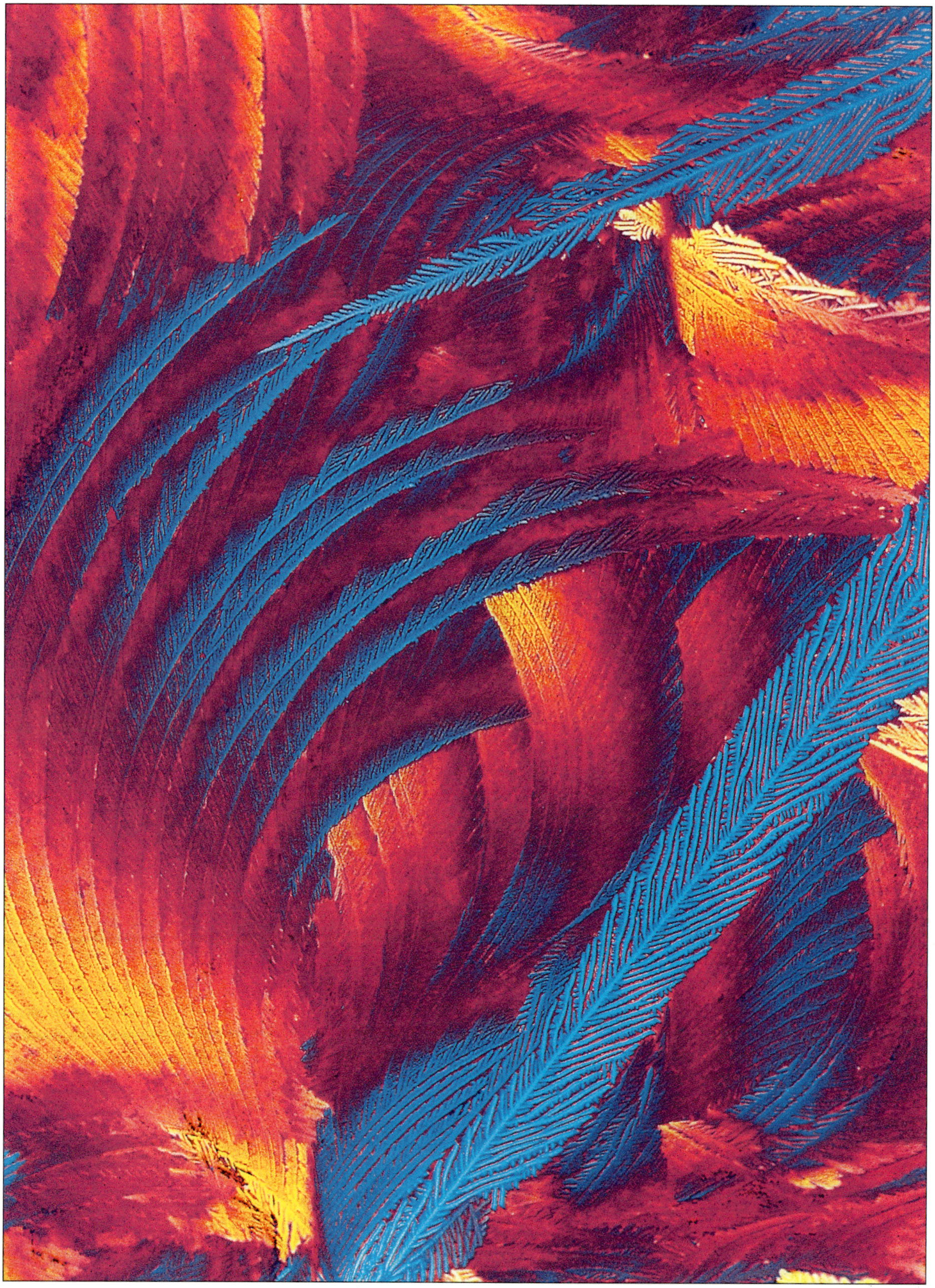

Alpha Tocopherol (Vitamin E)

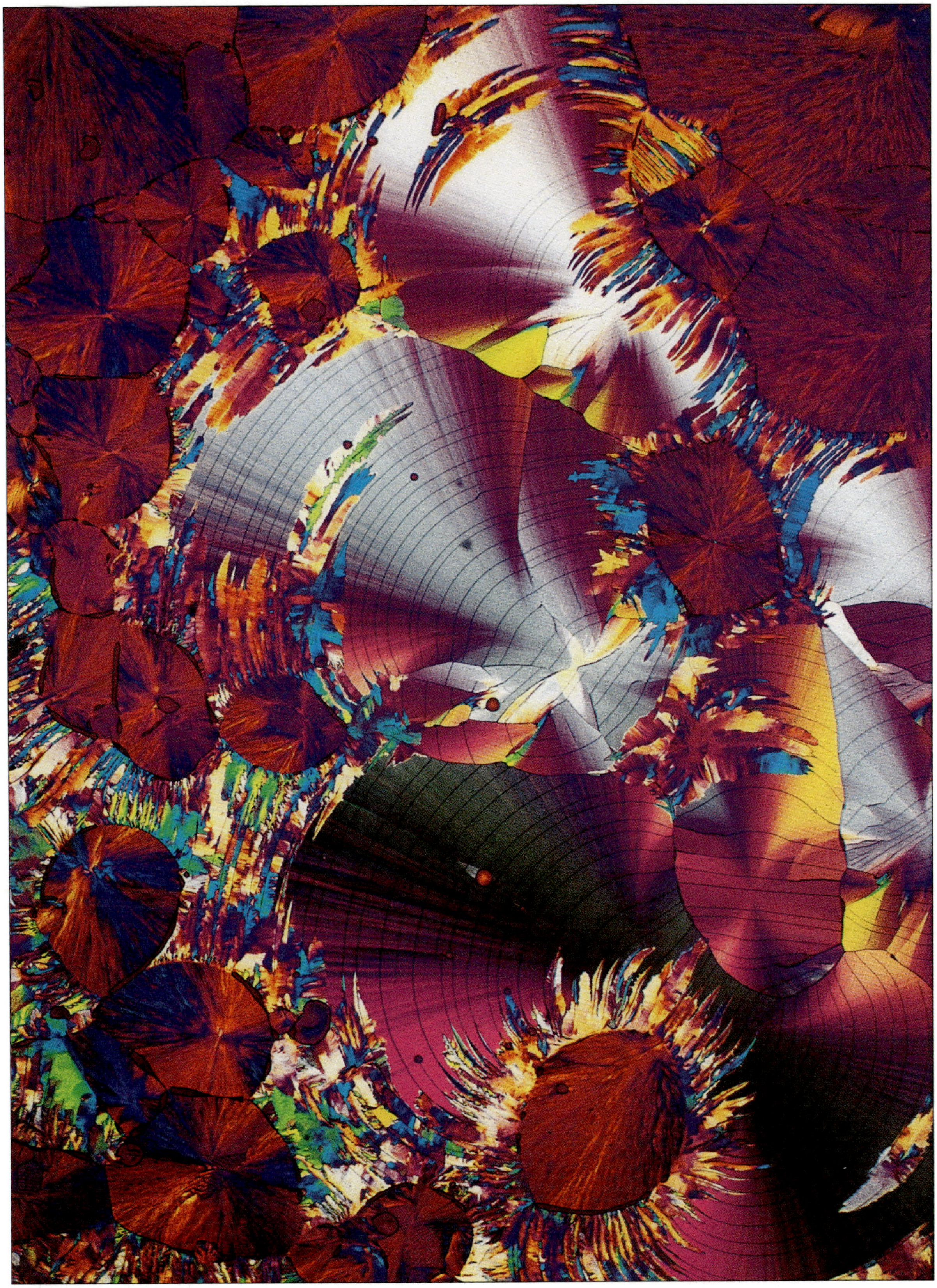

Retinoic Acid (Vitamin A)

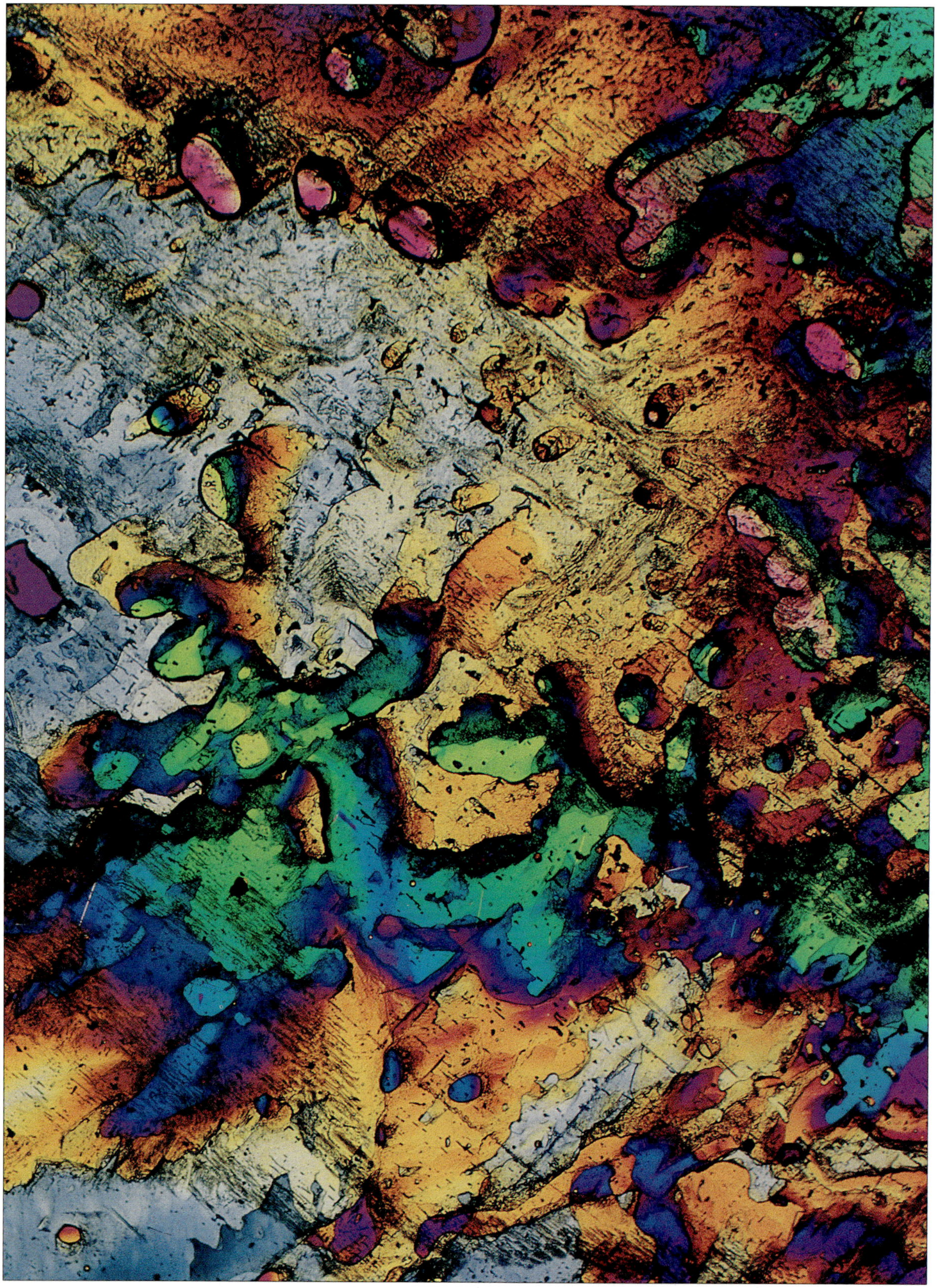

Ampicillin (Anti-Biotic)

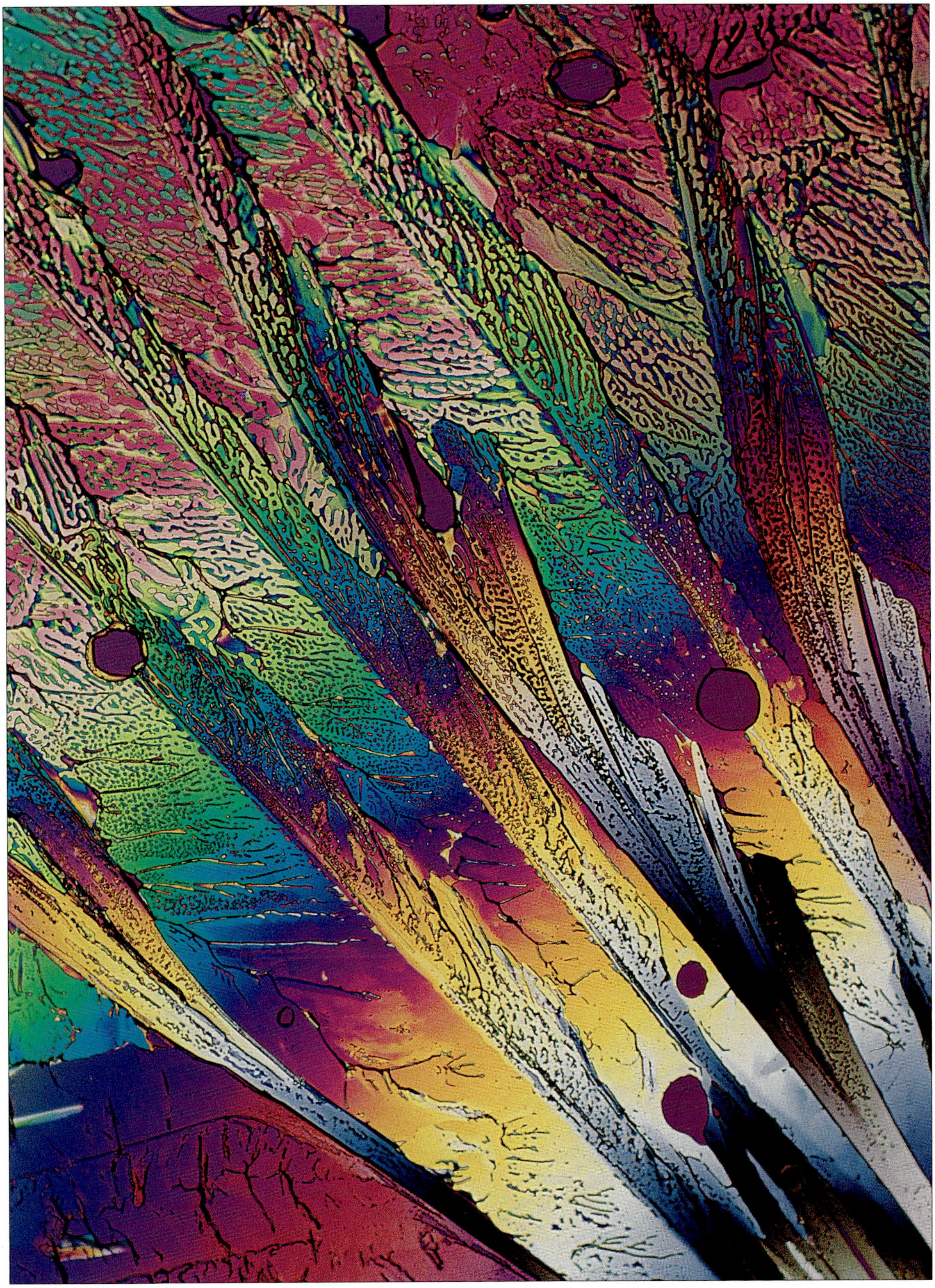

Cyanocobalamin (Vitamin B12)

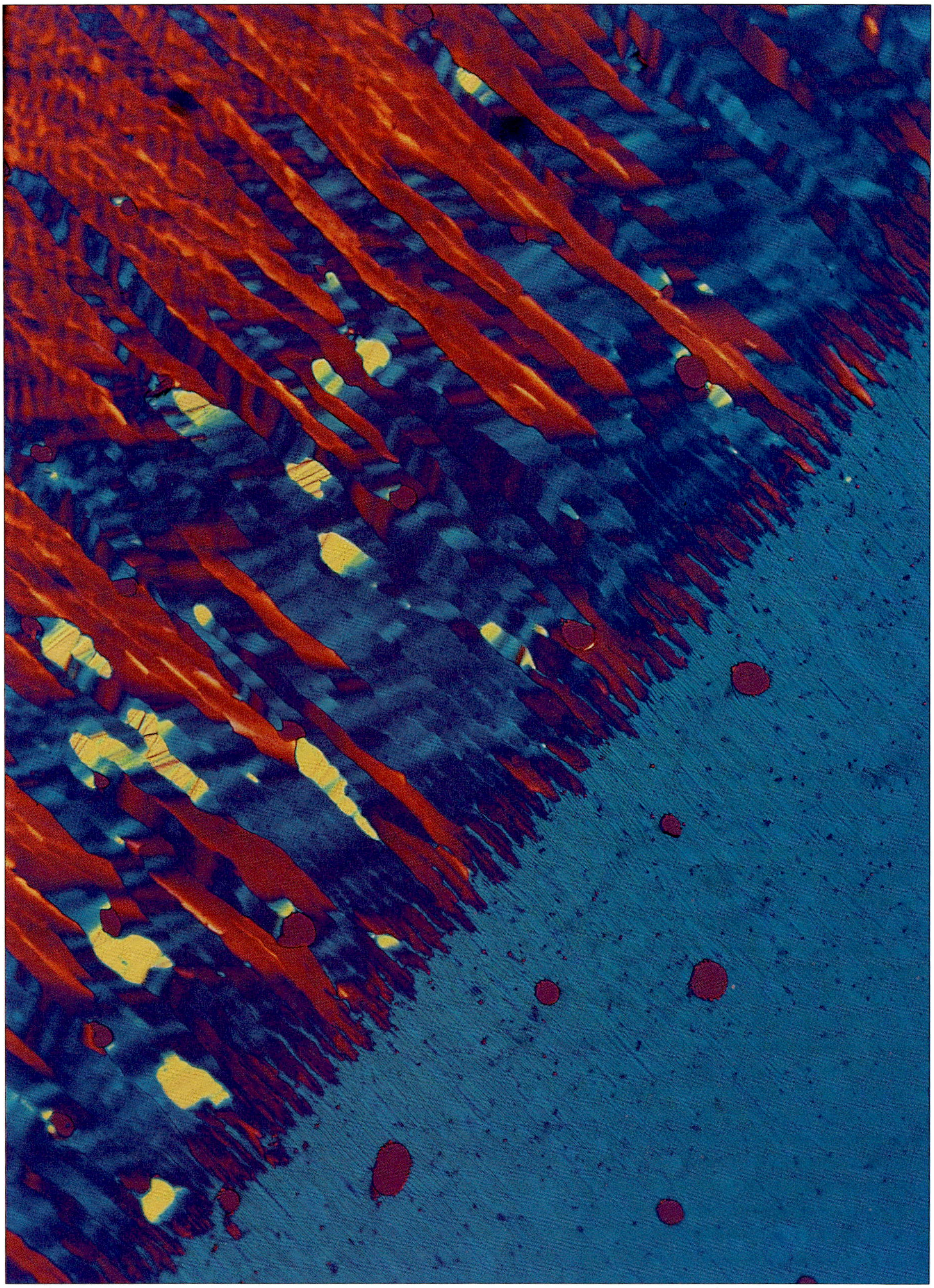

DNA

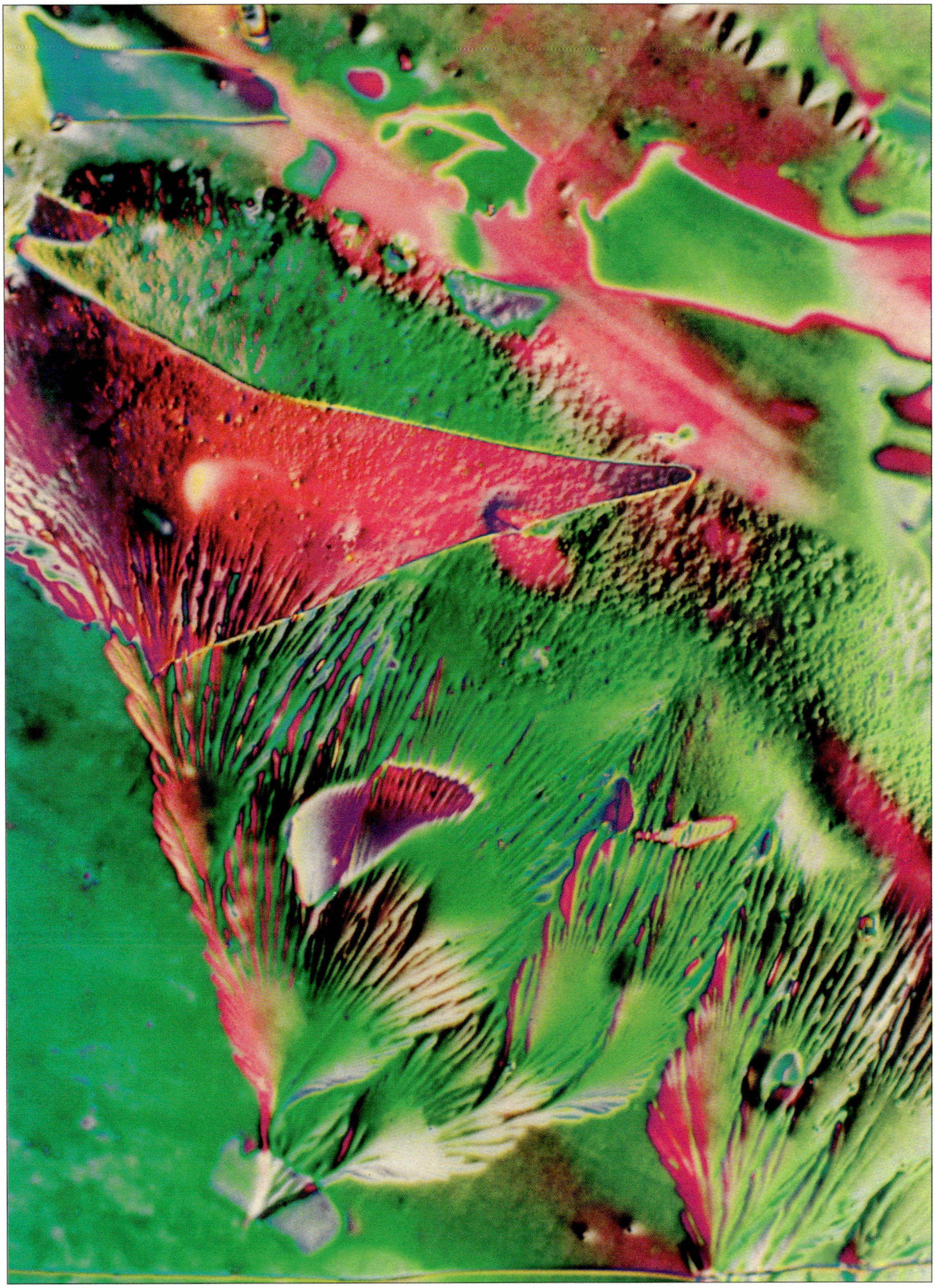

Liquid Crystal Display

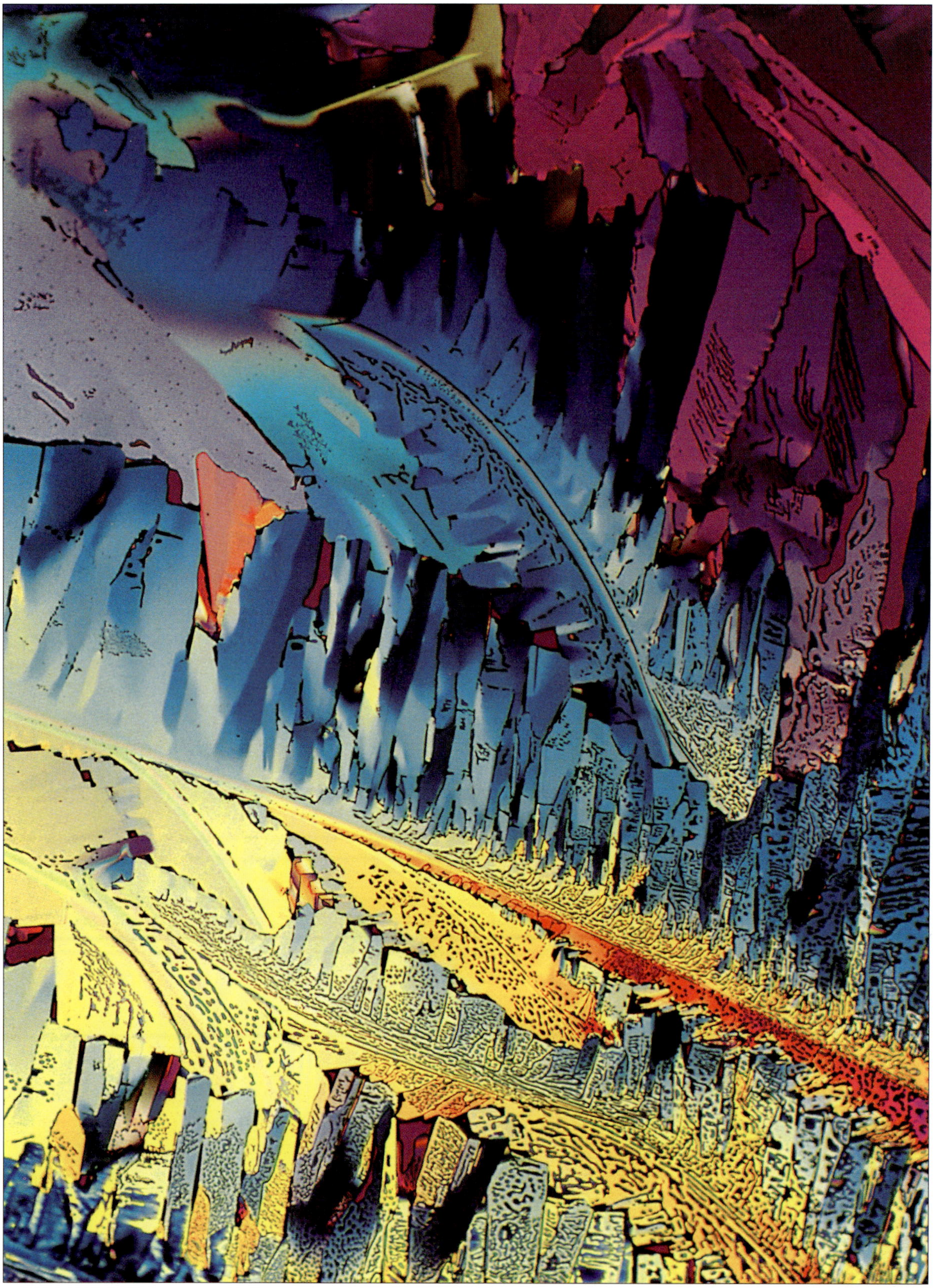

Orotic Acid (Vitamin B13)

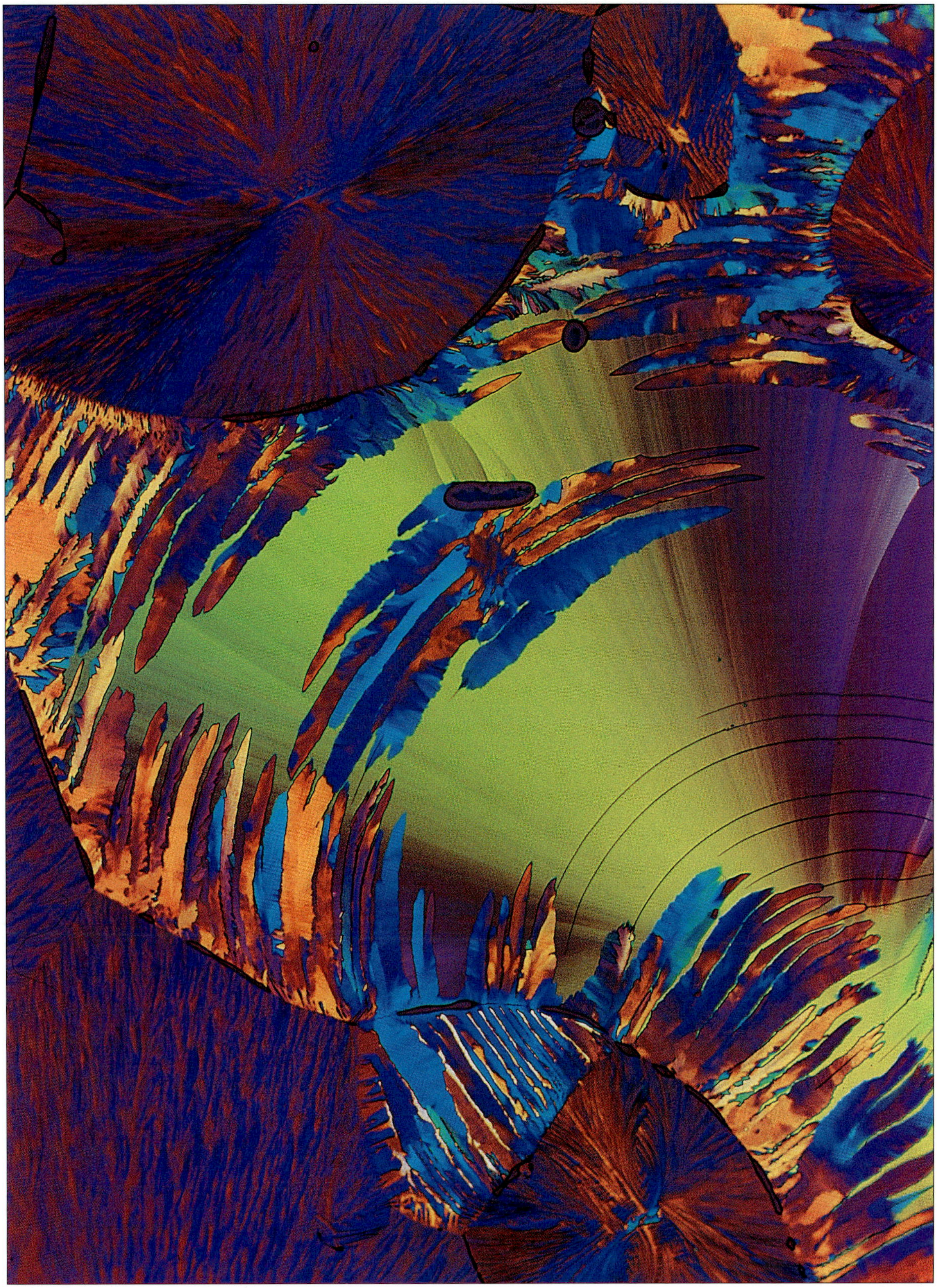
Retinoic Acid (Vitamin A)

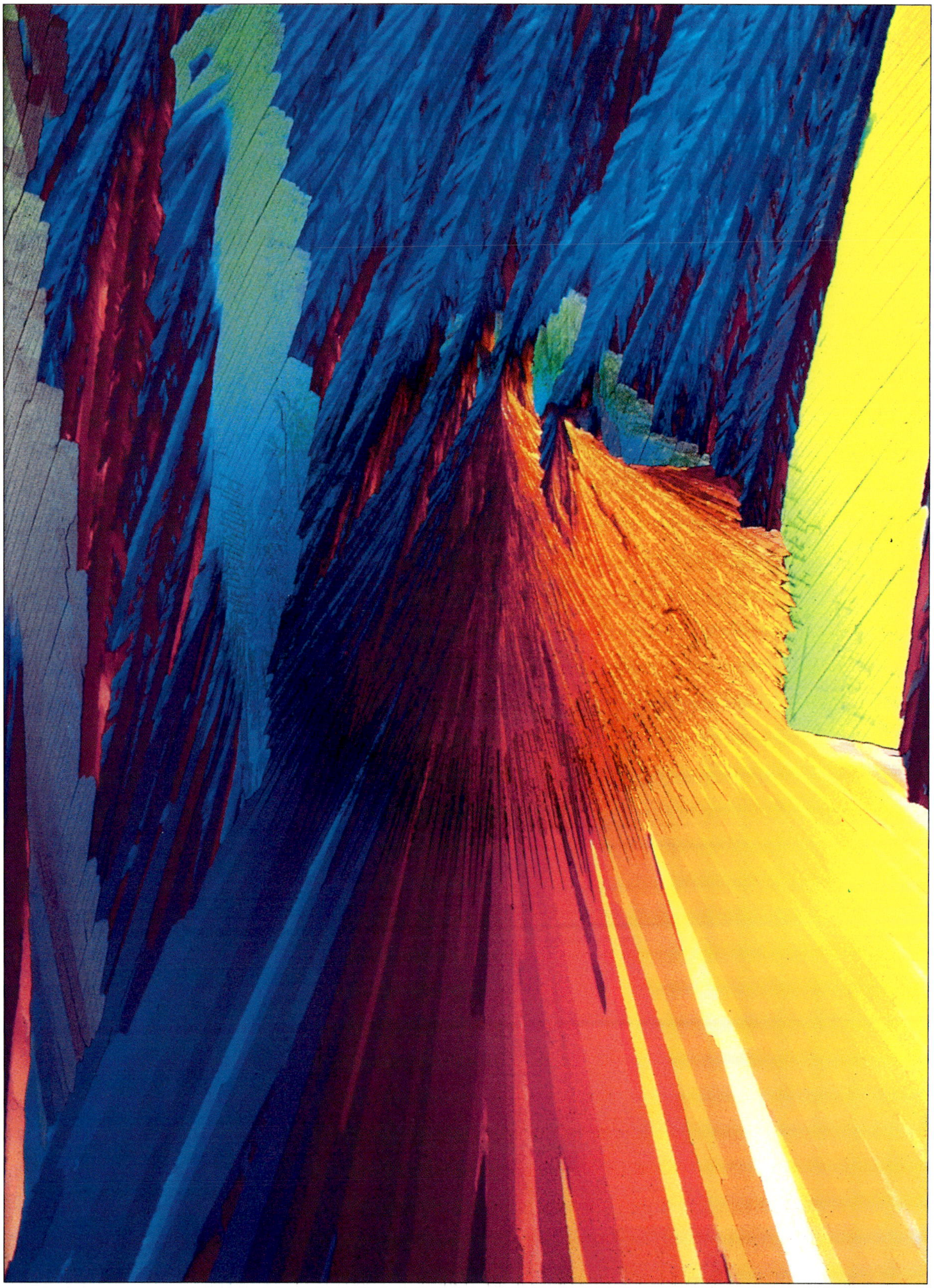

Thiamine (Vitamin B1)

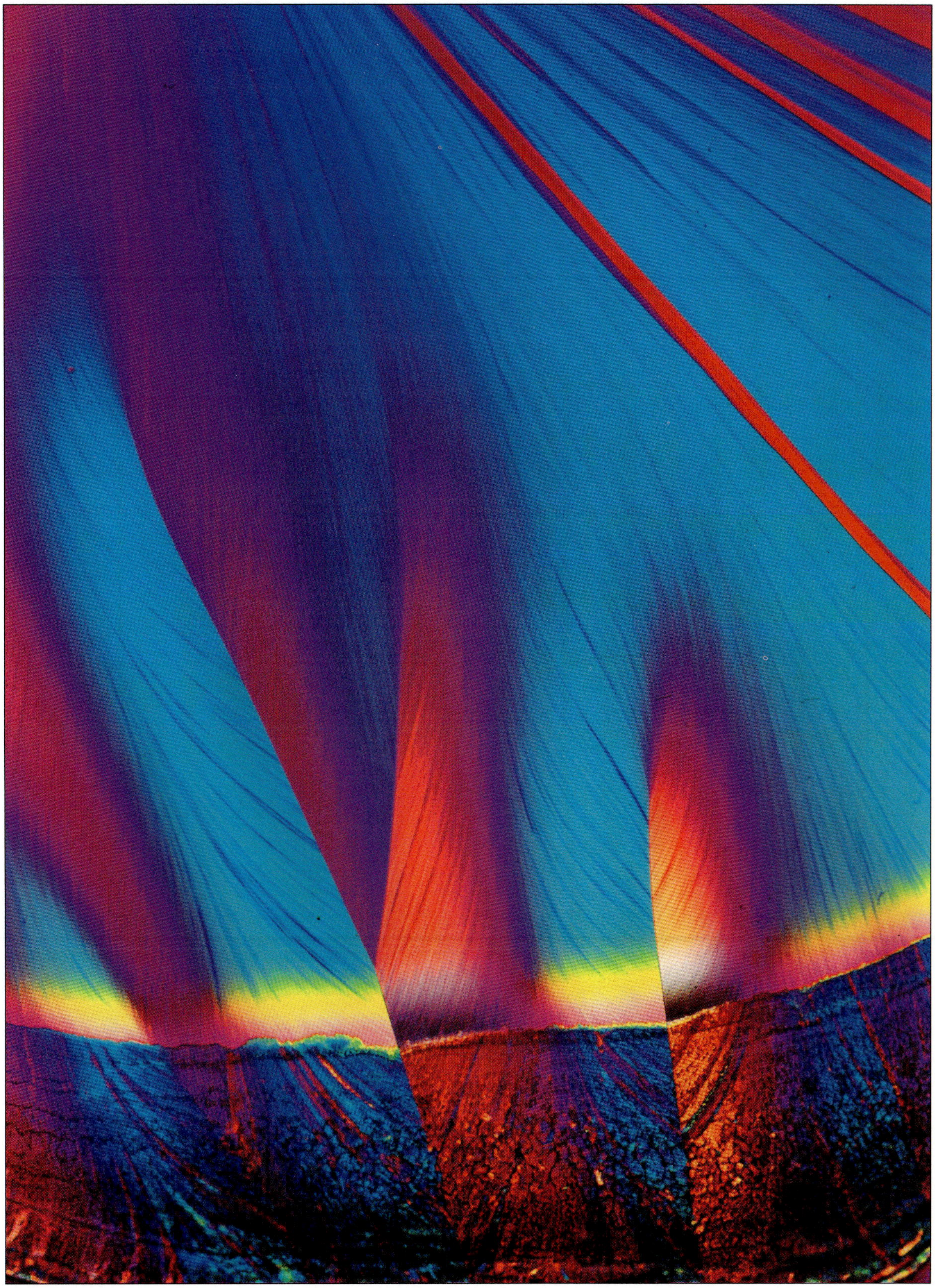
DDC (Anti-AIDS)

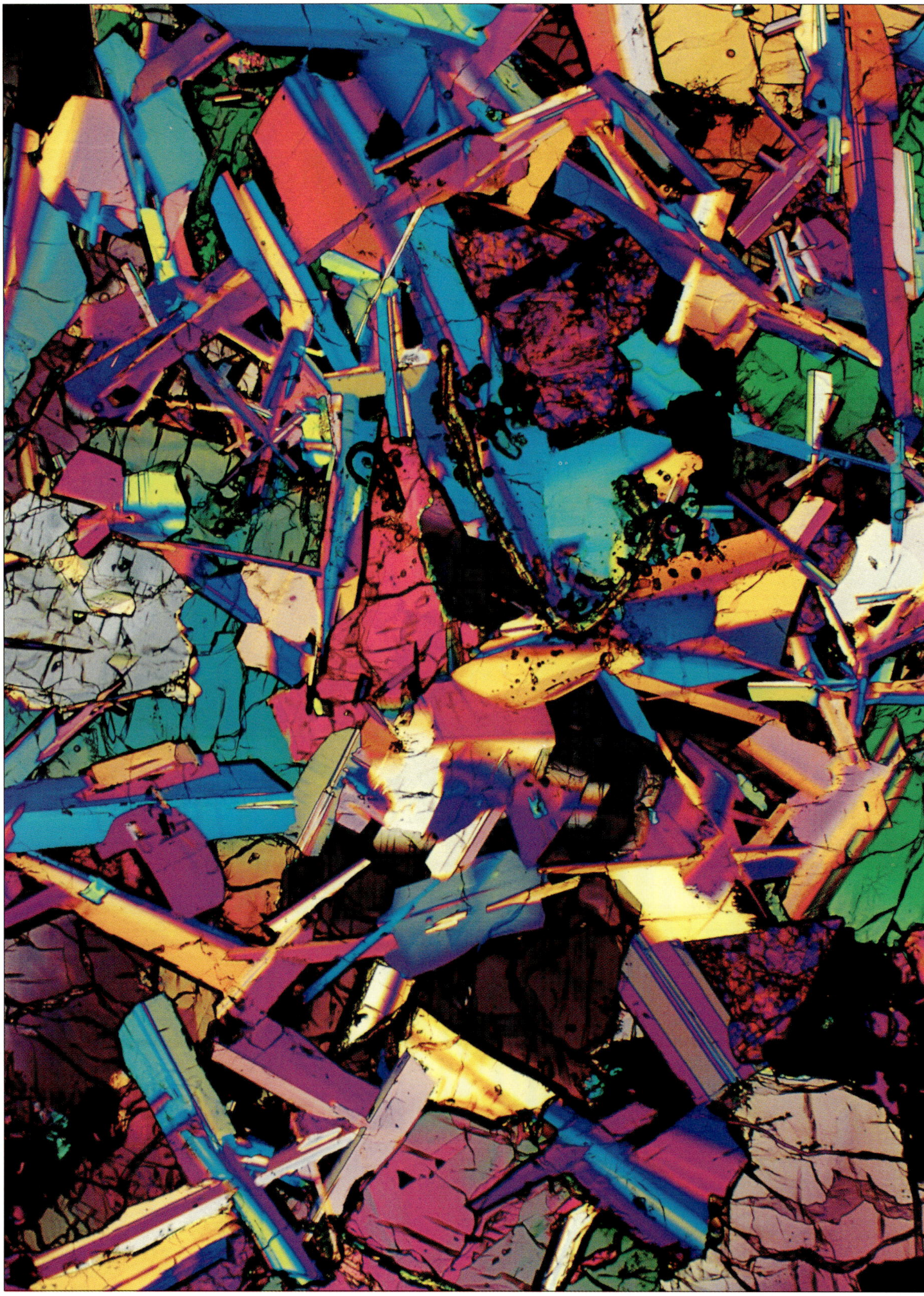

Apollo 11 Moon Rock

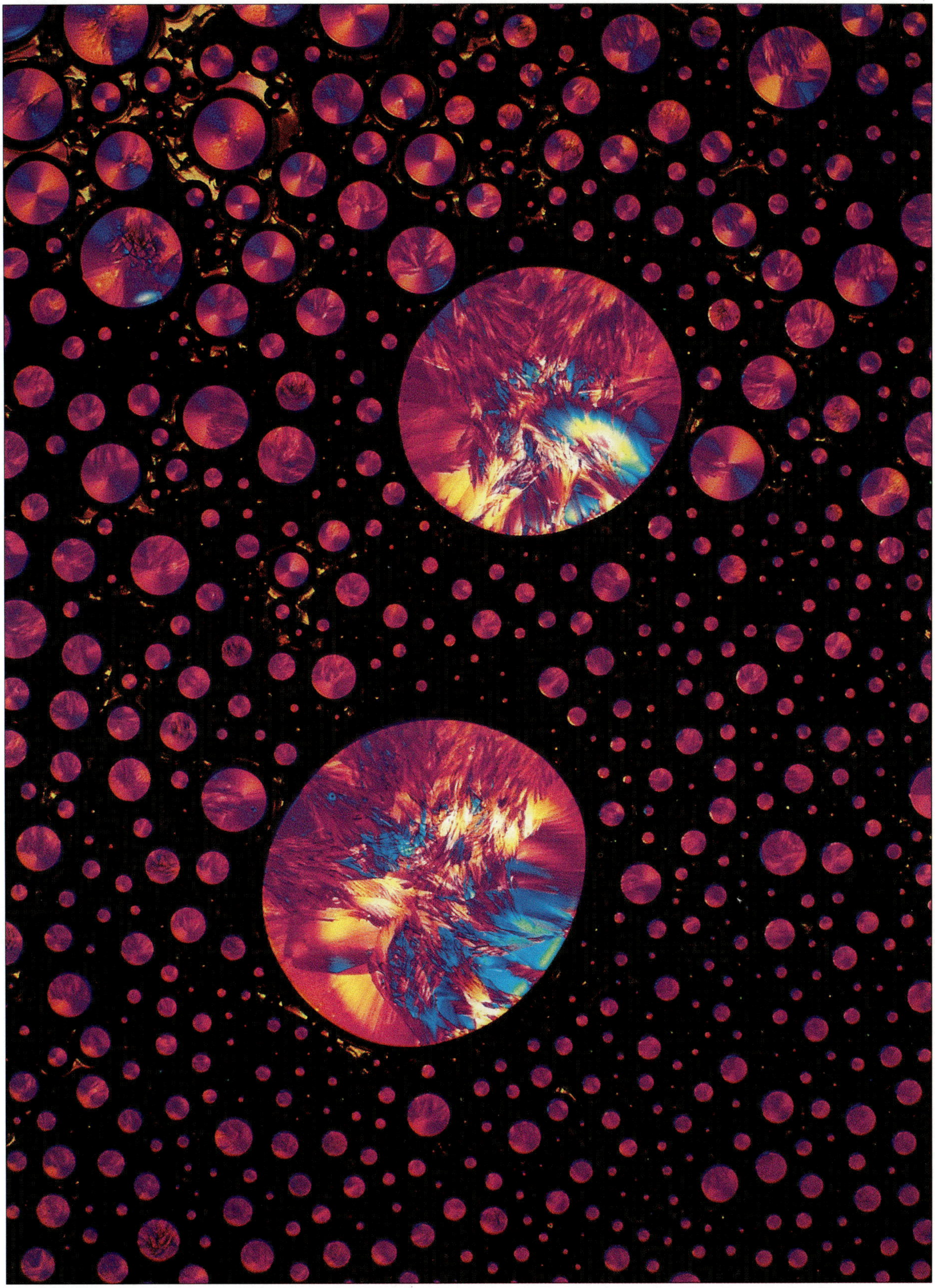

Ascorbic Acid (Vitamin C)

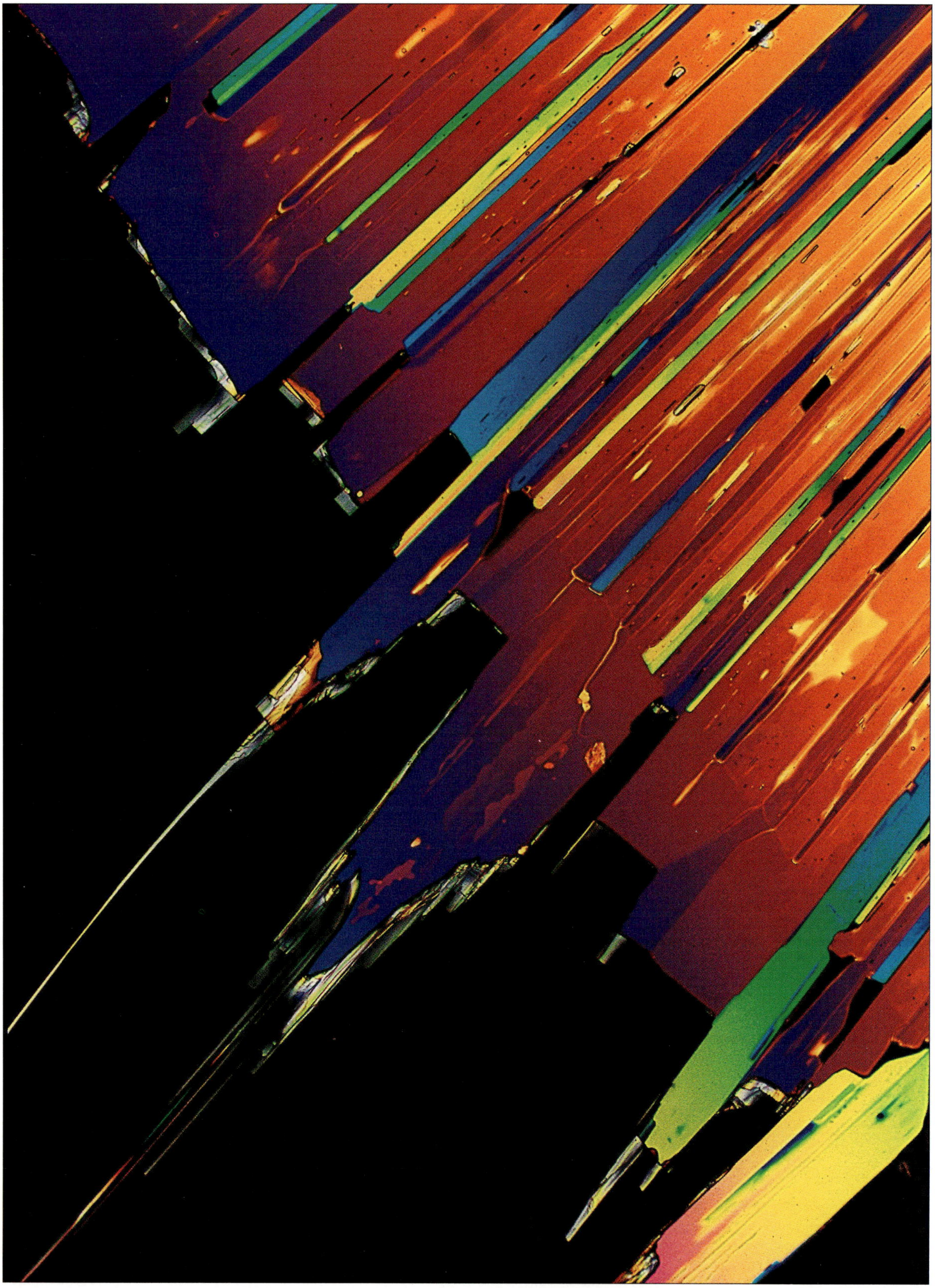

Cocaine HCl

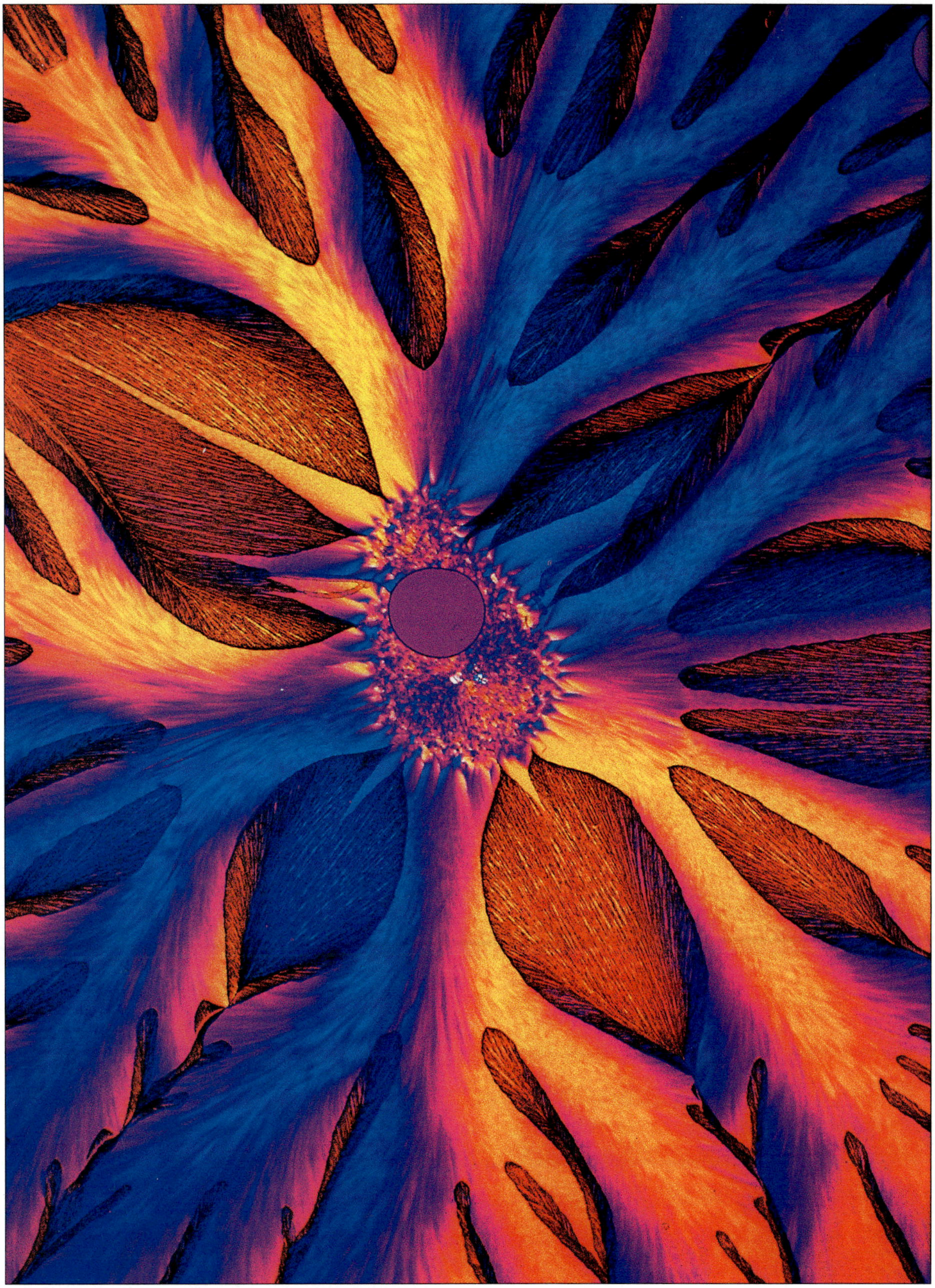

Alpha Tocopherol (Vitamin E)

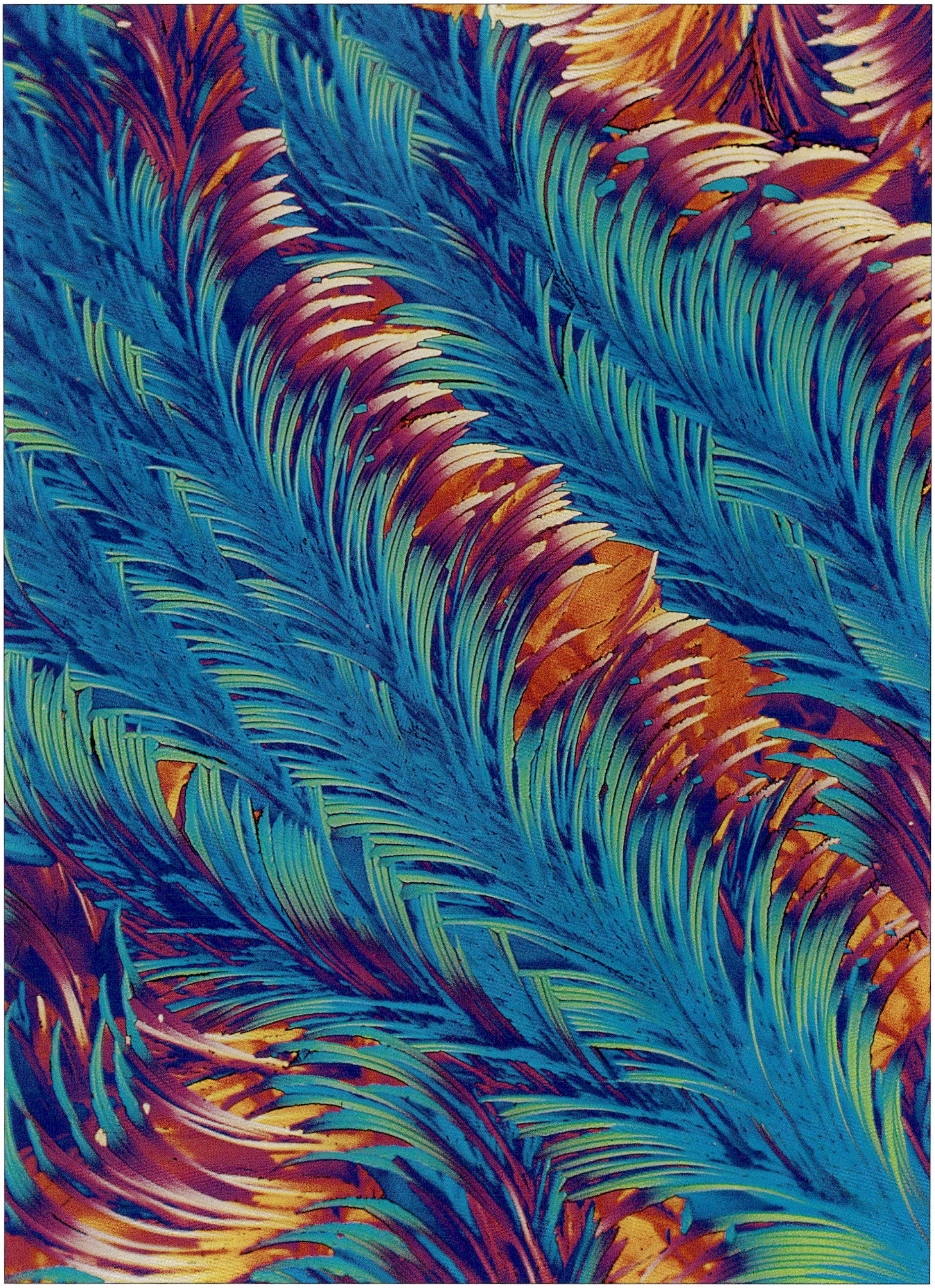

Riboflavin (Vitamin B2)

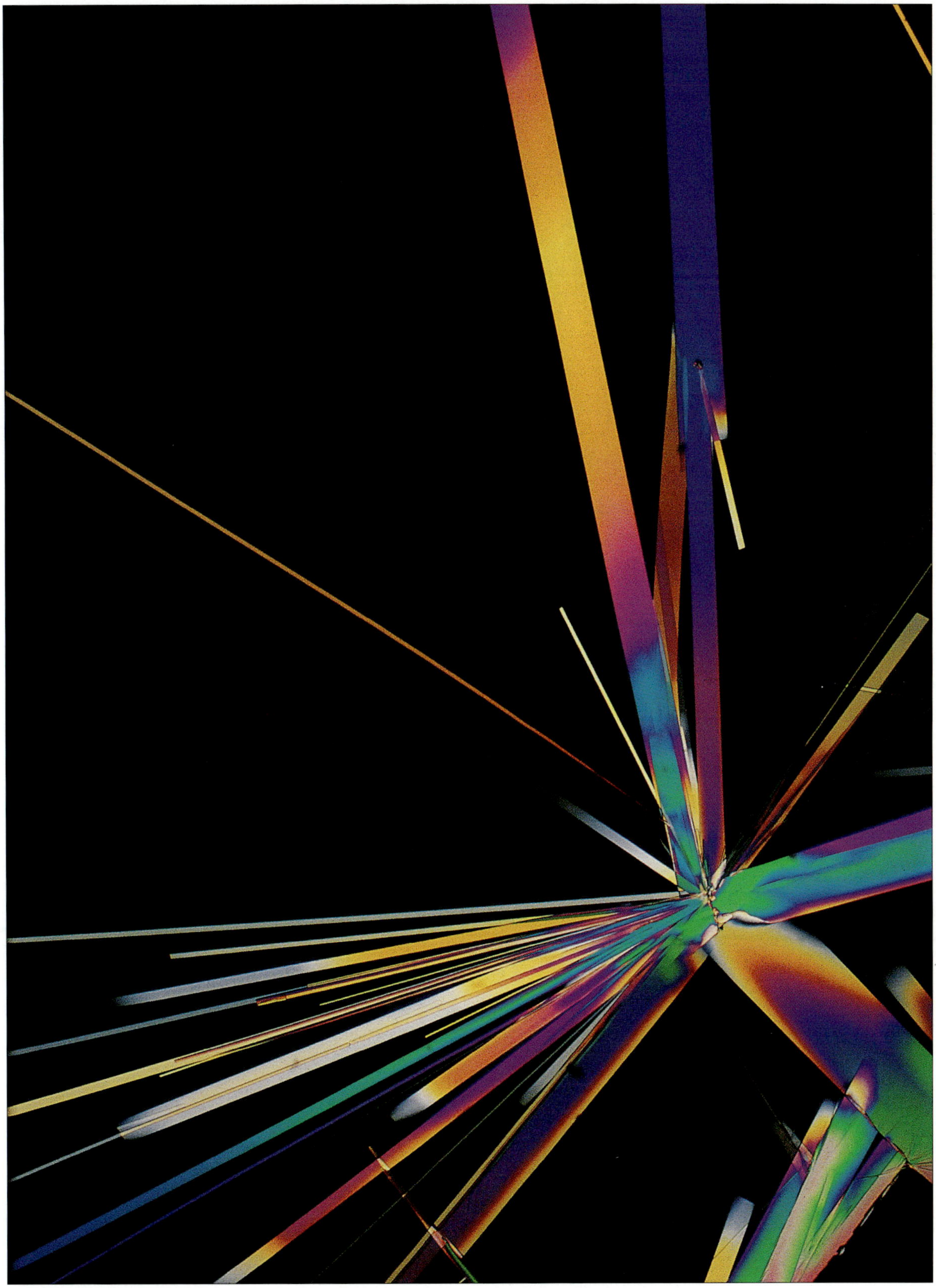

Thiamine (Vitamin B1)

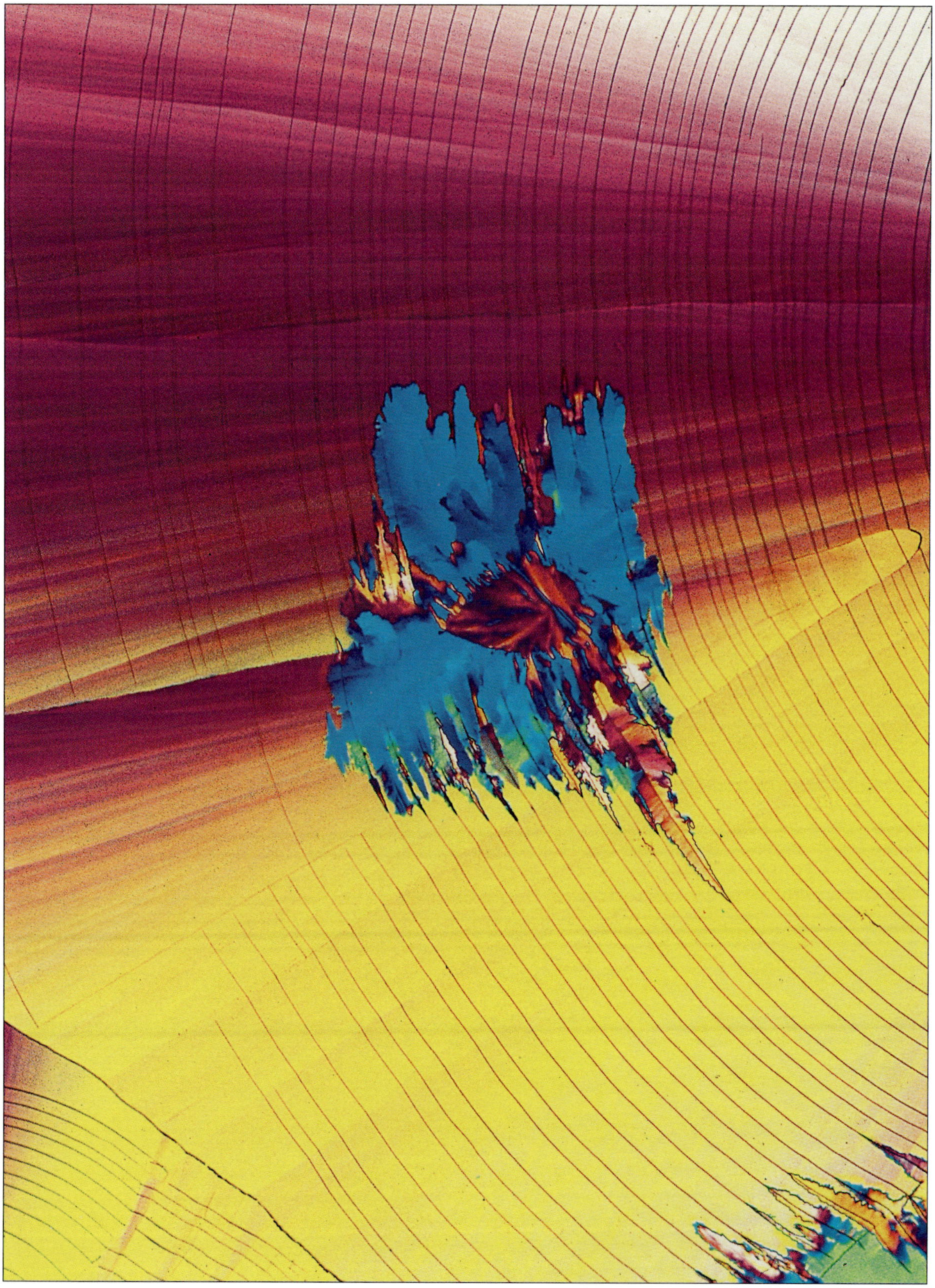
Retinoic Acid (Vitamin A)

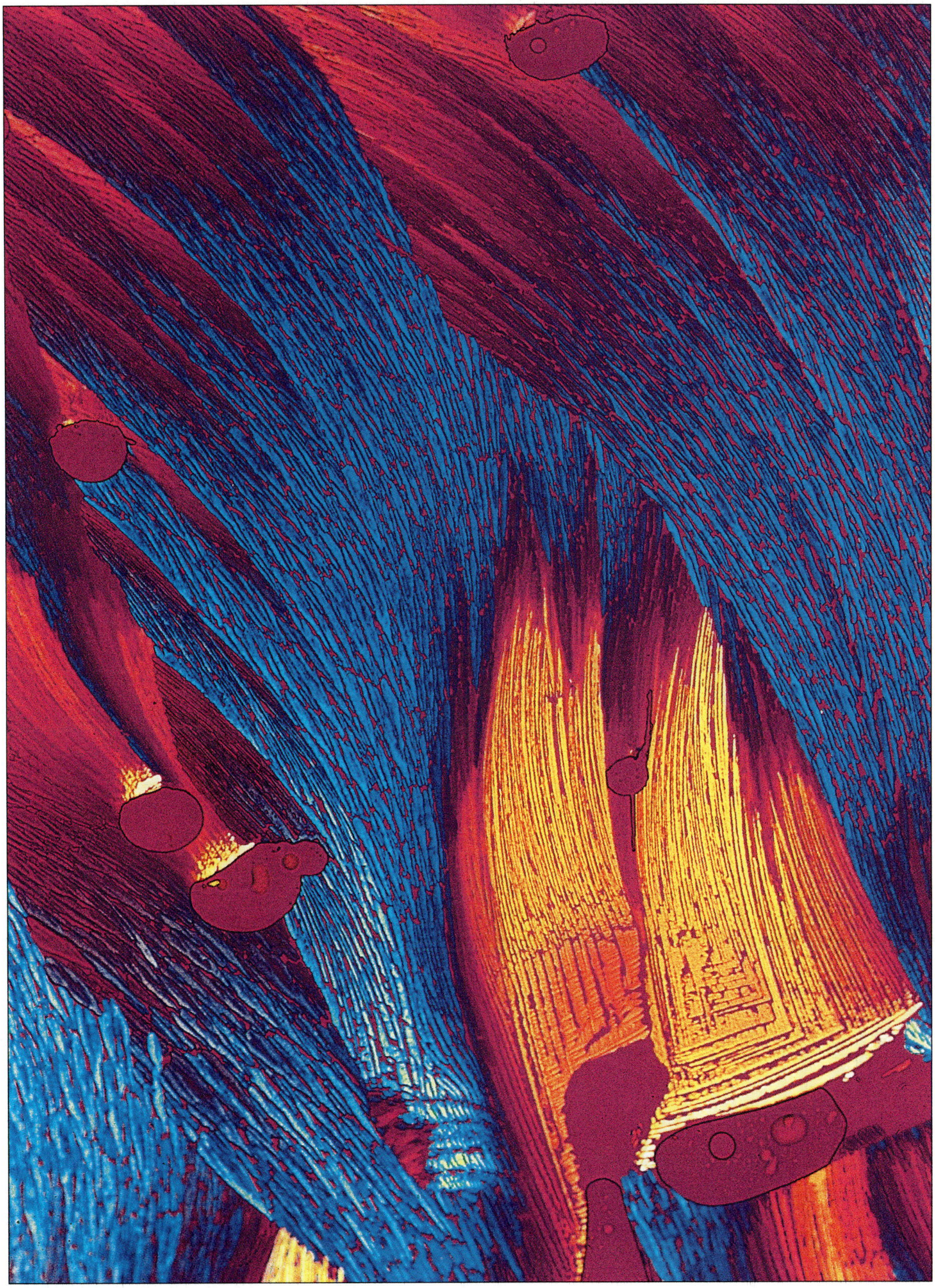
Alpha Tocopherol (Vitamin E)

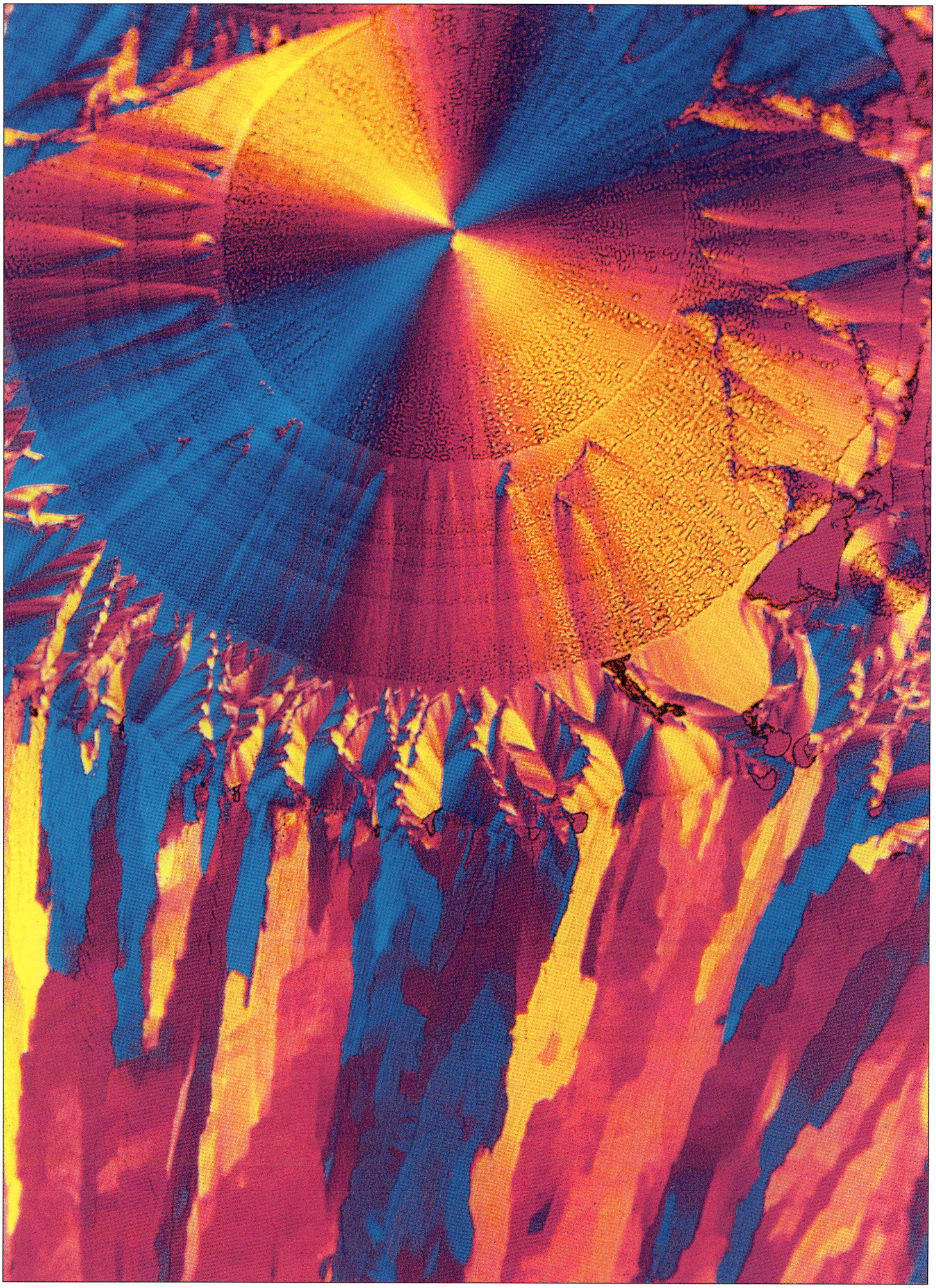

Erythromycin

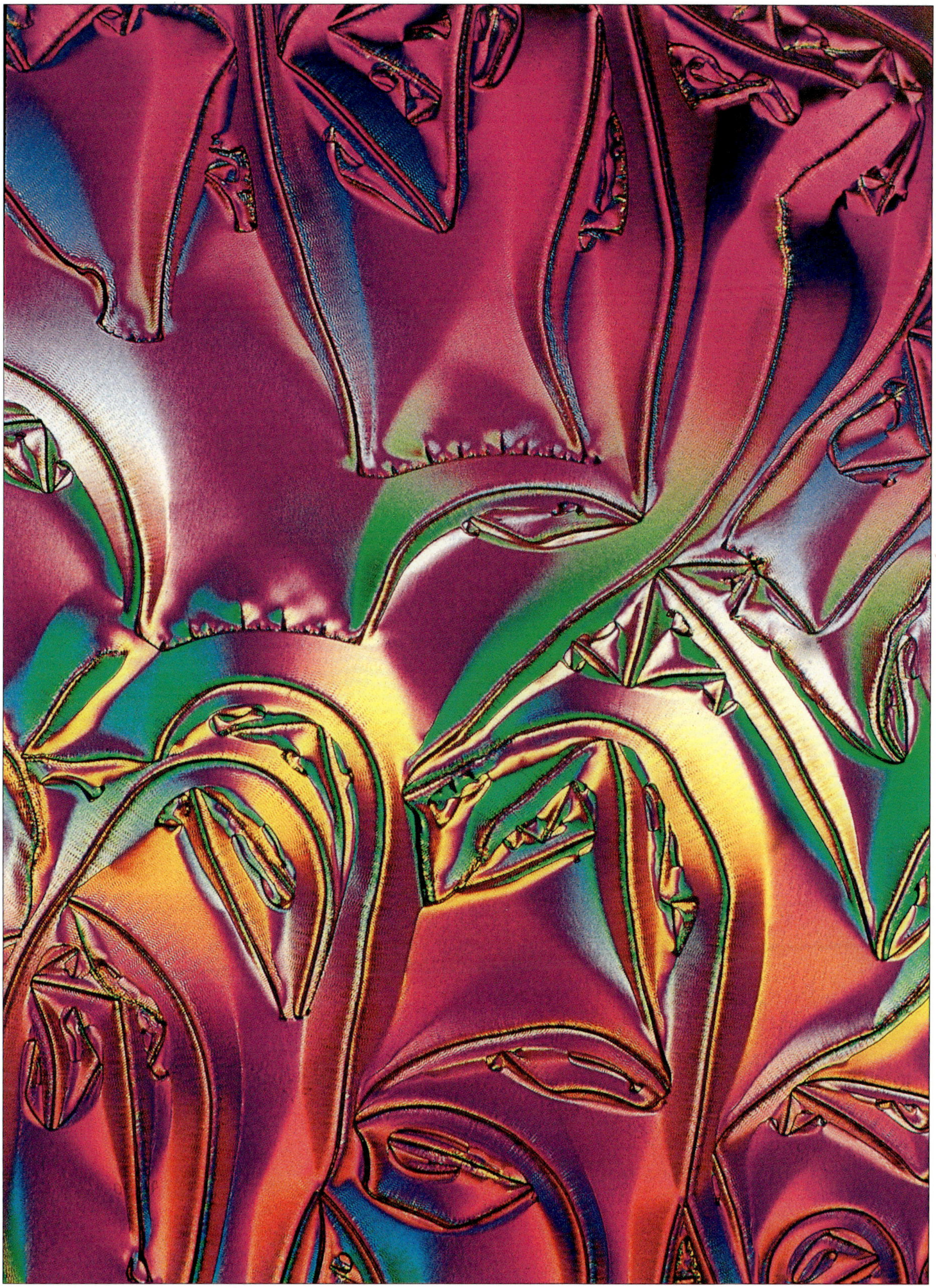

Ascorbic Acid (Vitamin C)

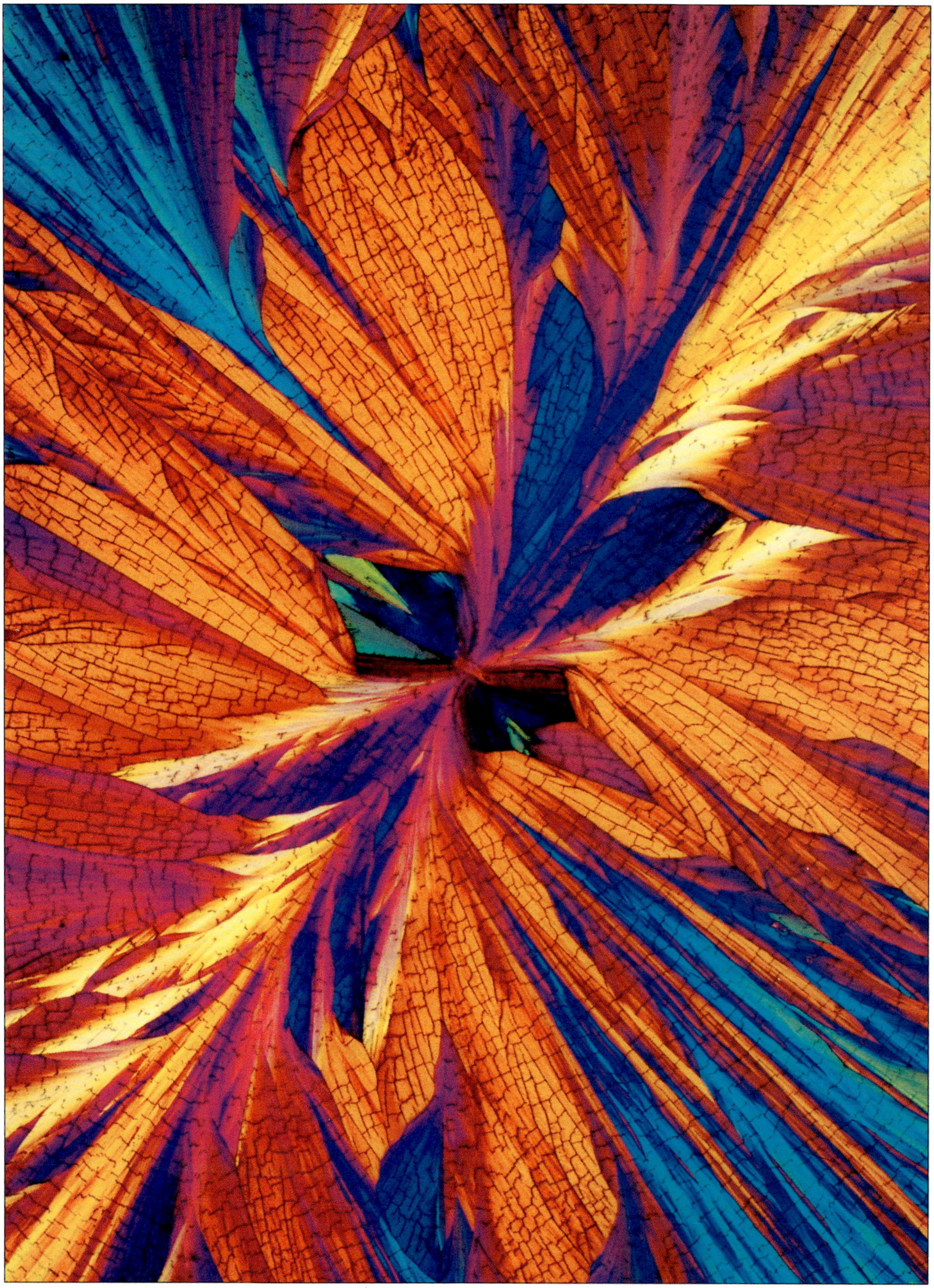

Lemon

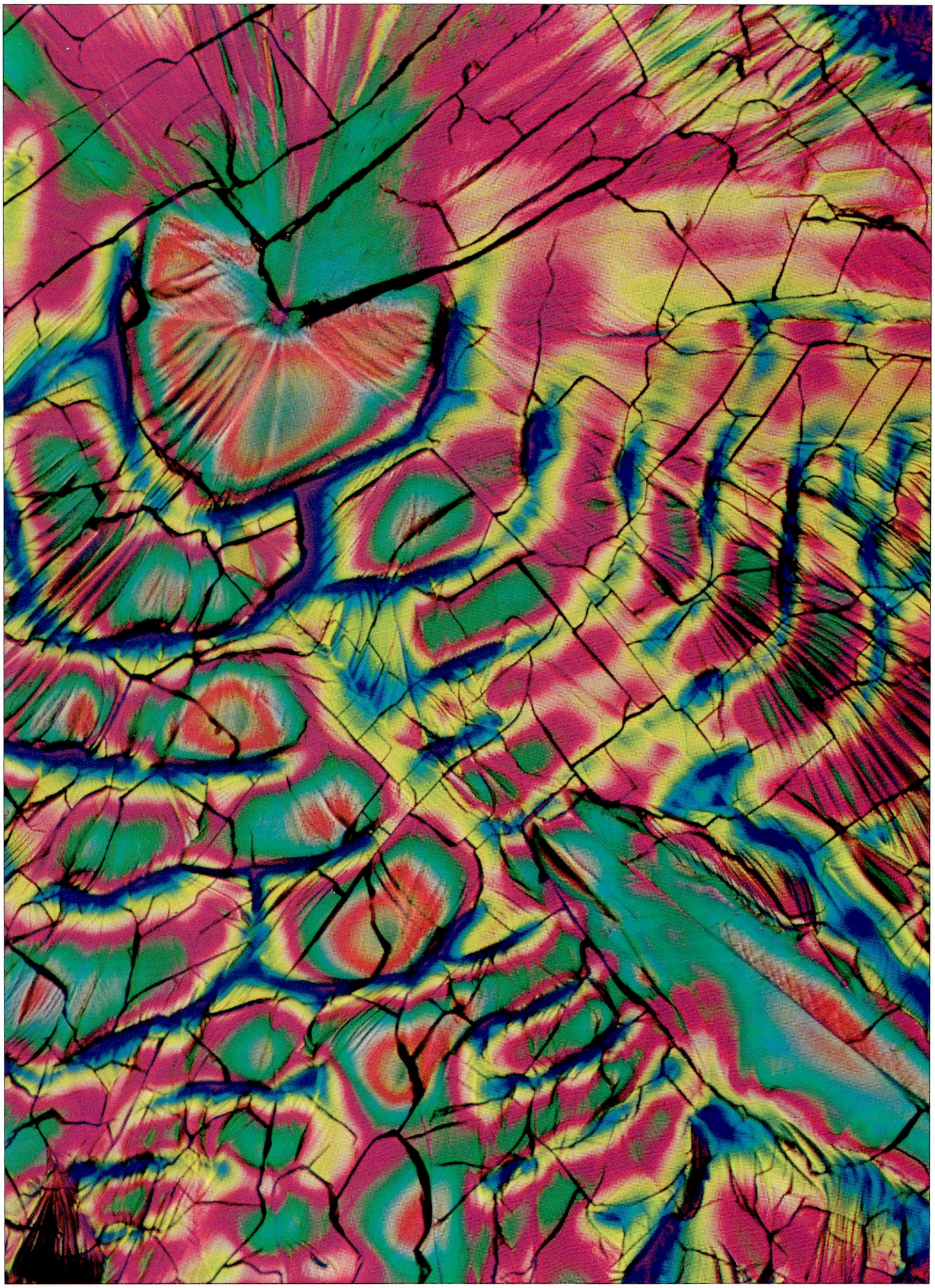
Watermelon

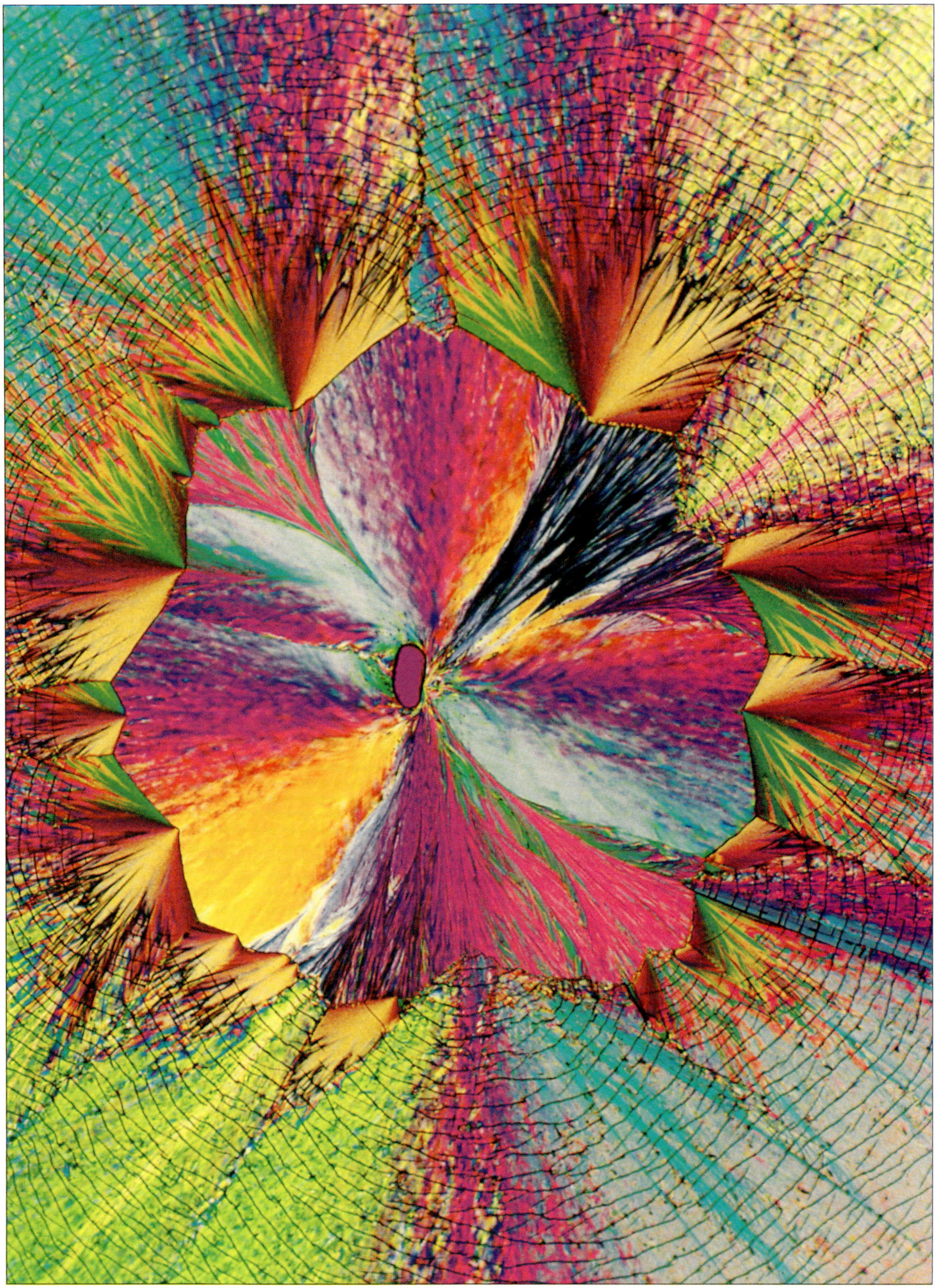

Popcorn

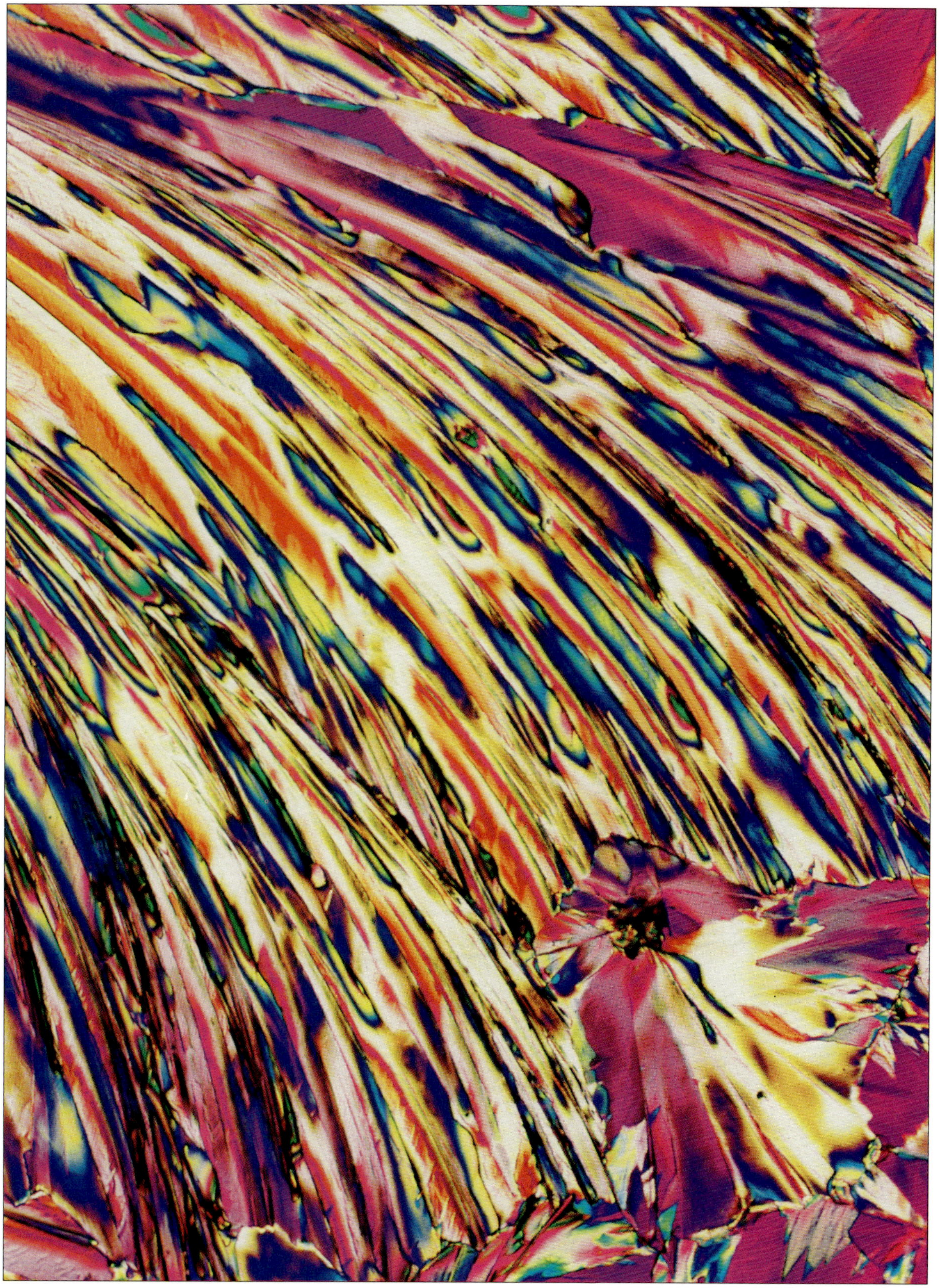

Grape

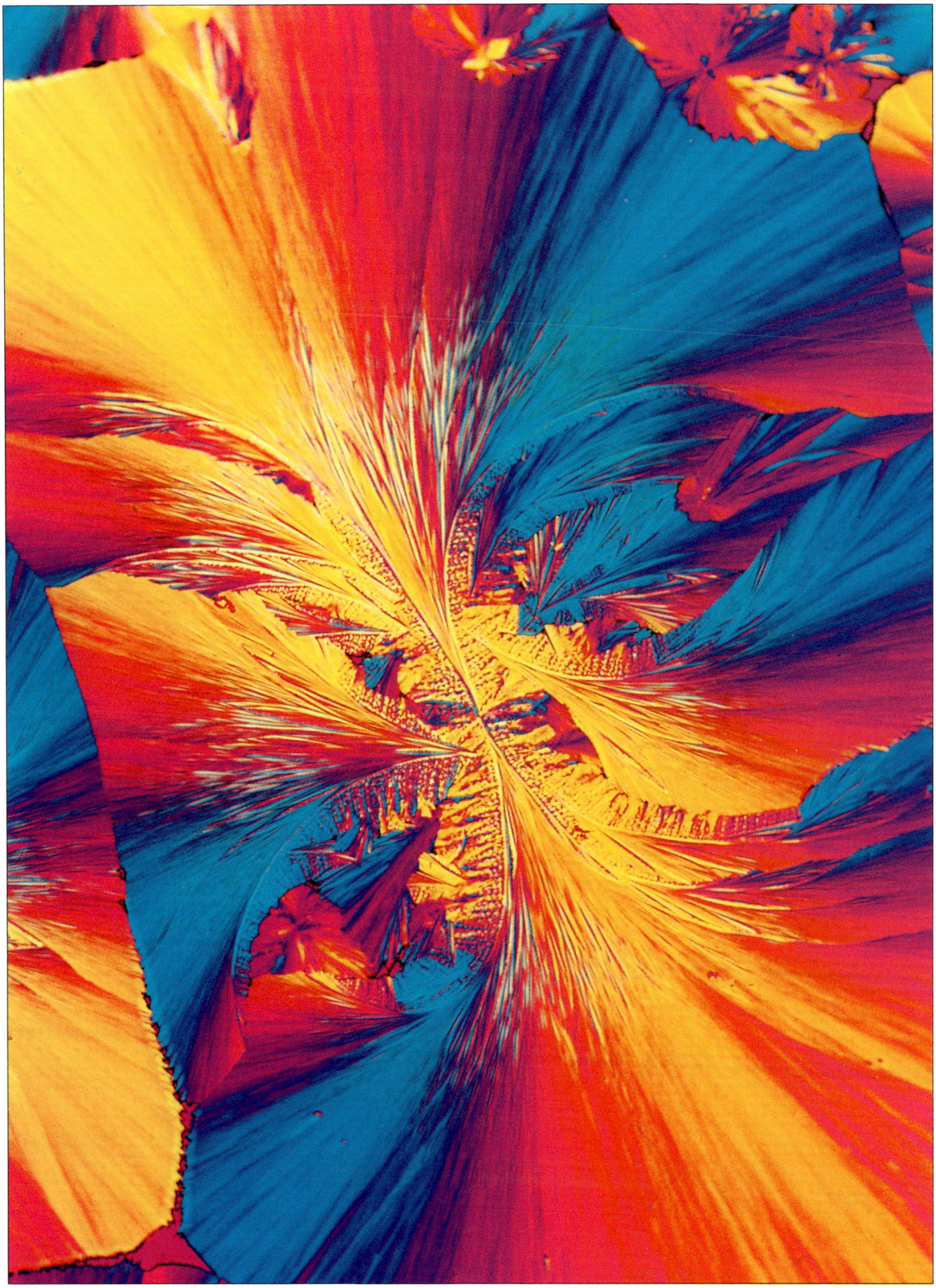

Prostaglandin

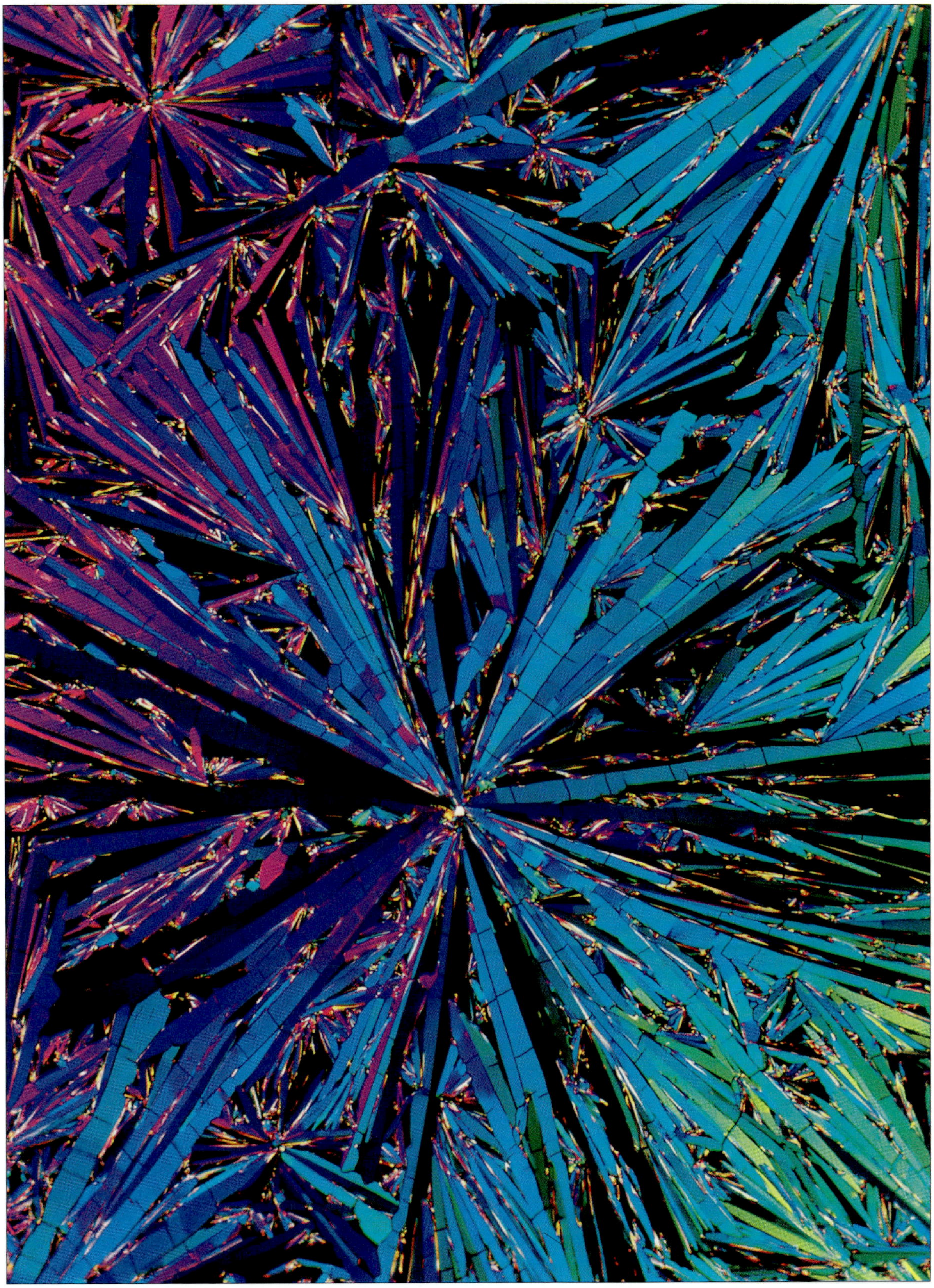
Lotrimin

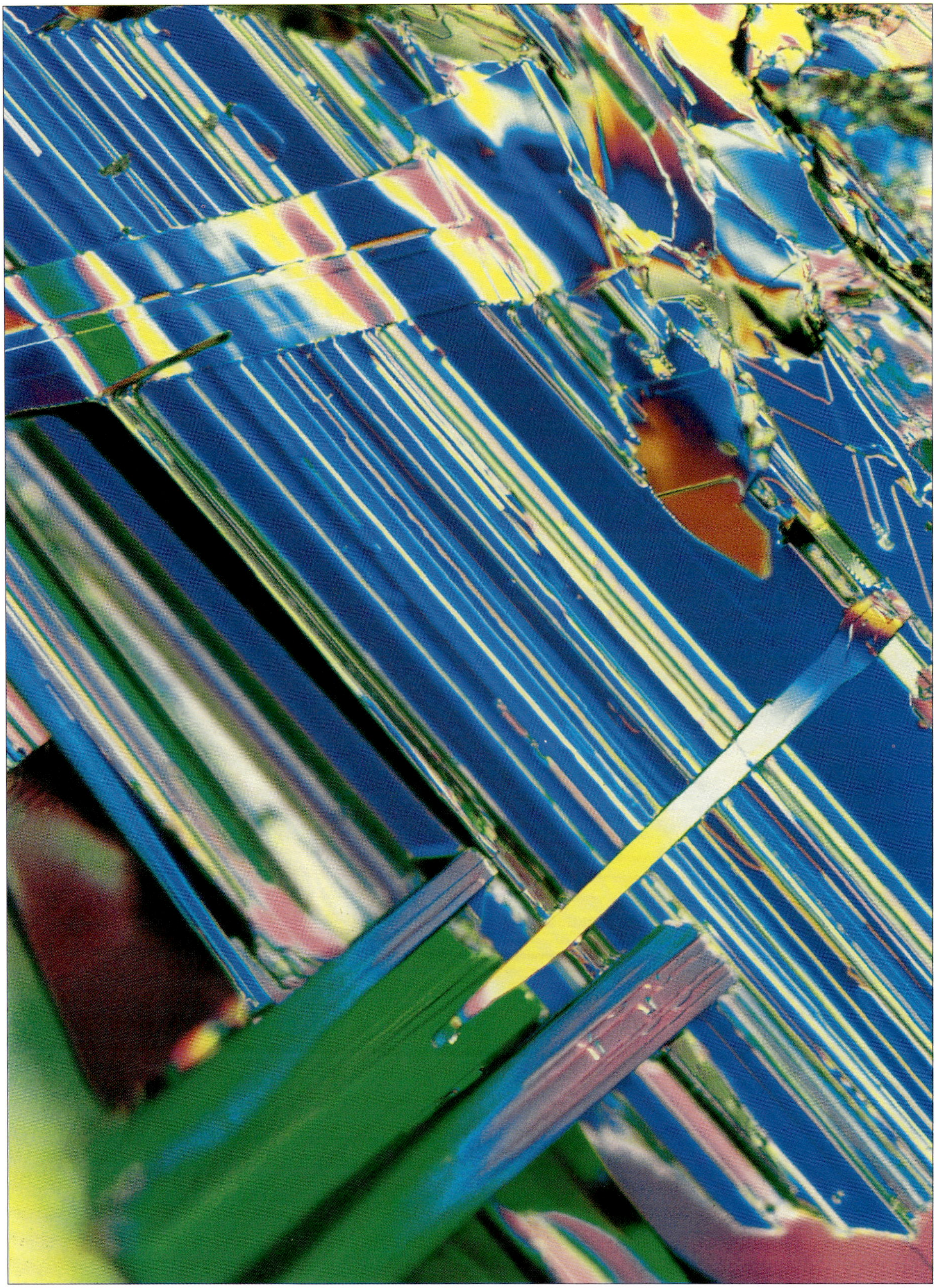

Superconductor

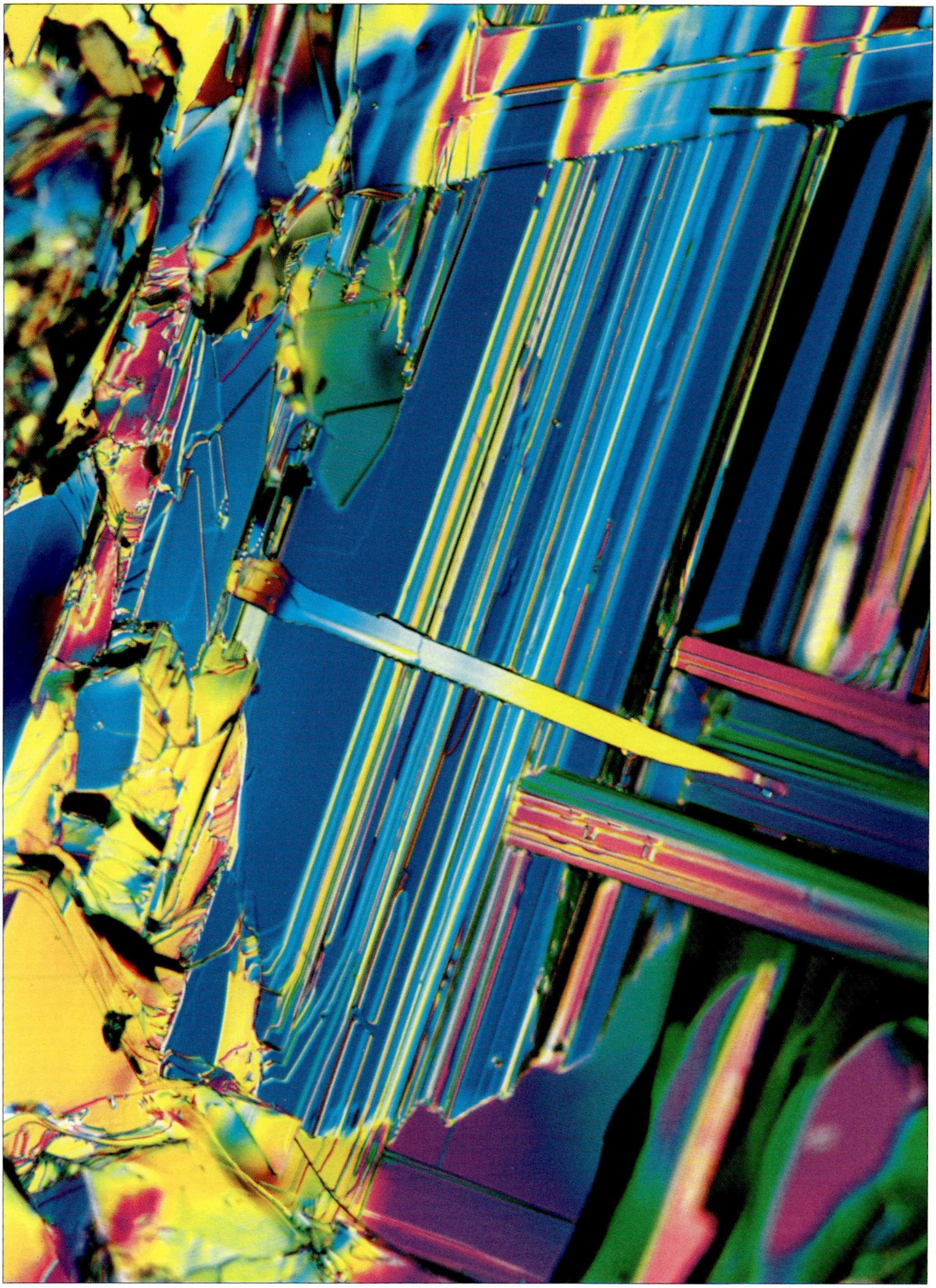

Single Crystal Surface

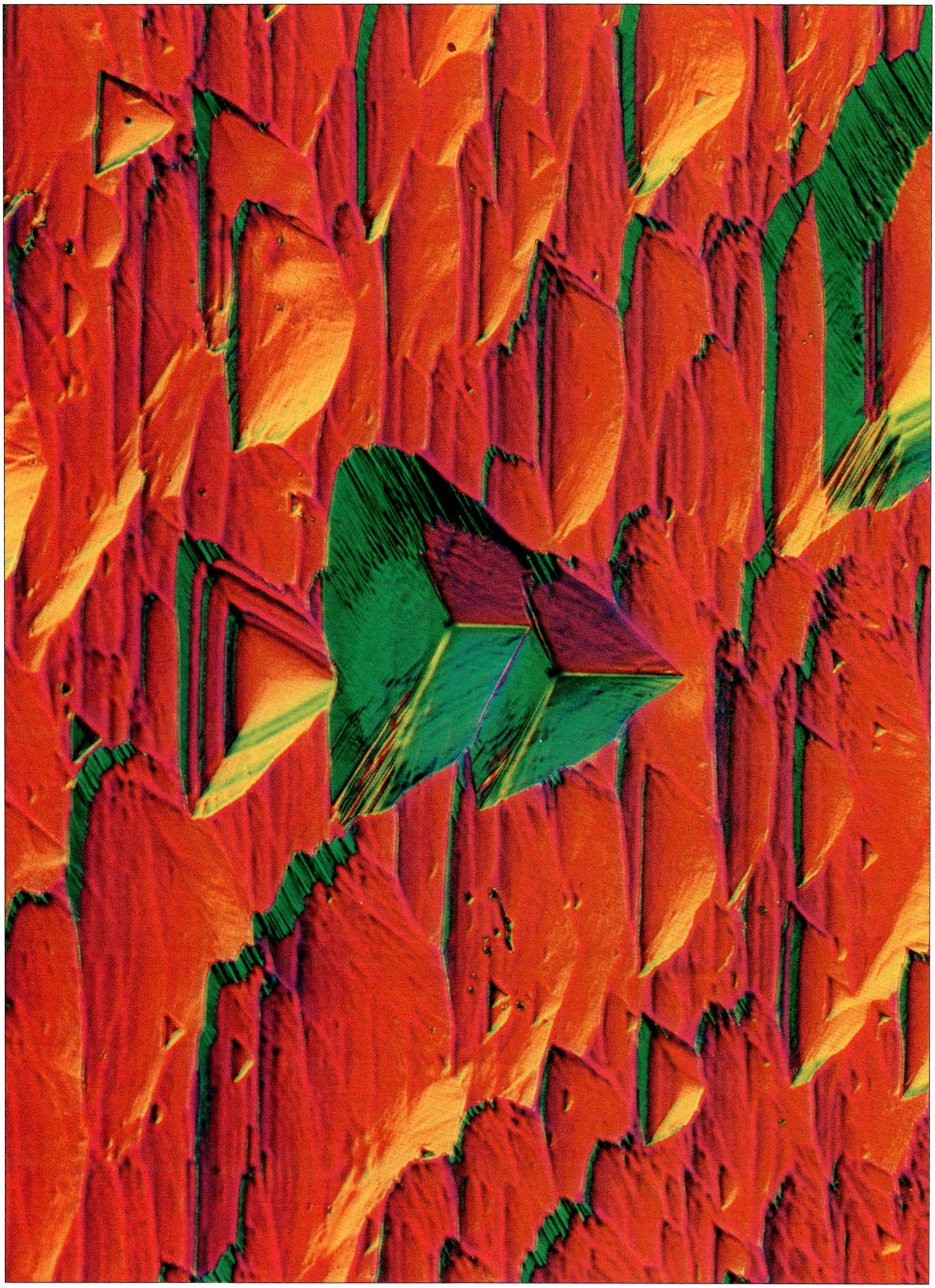

Silver

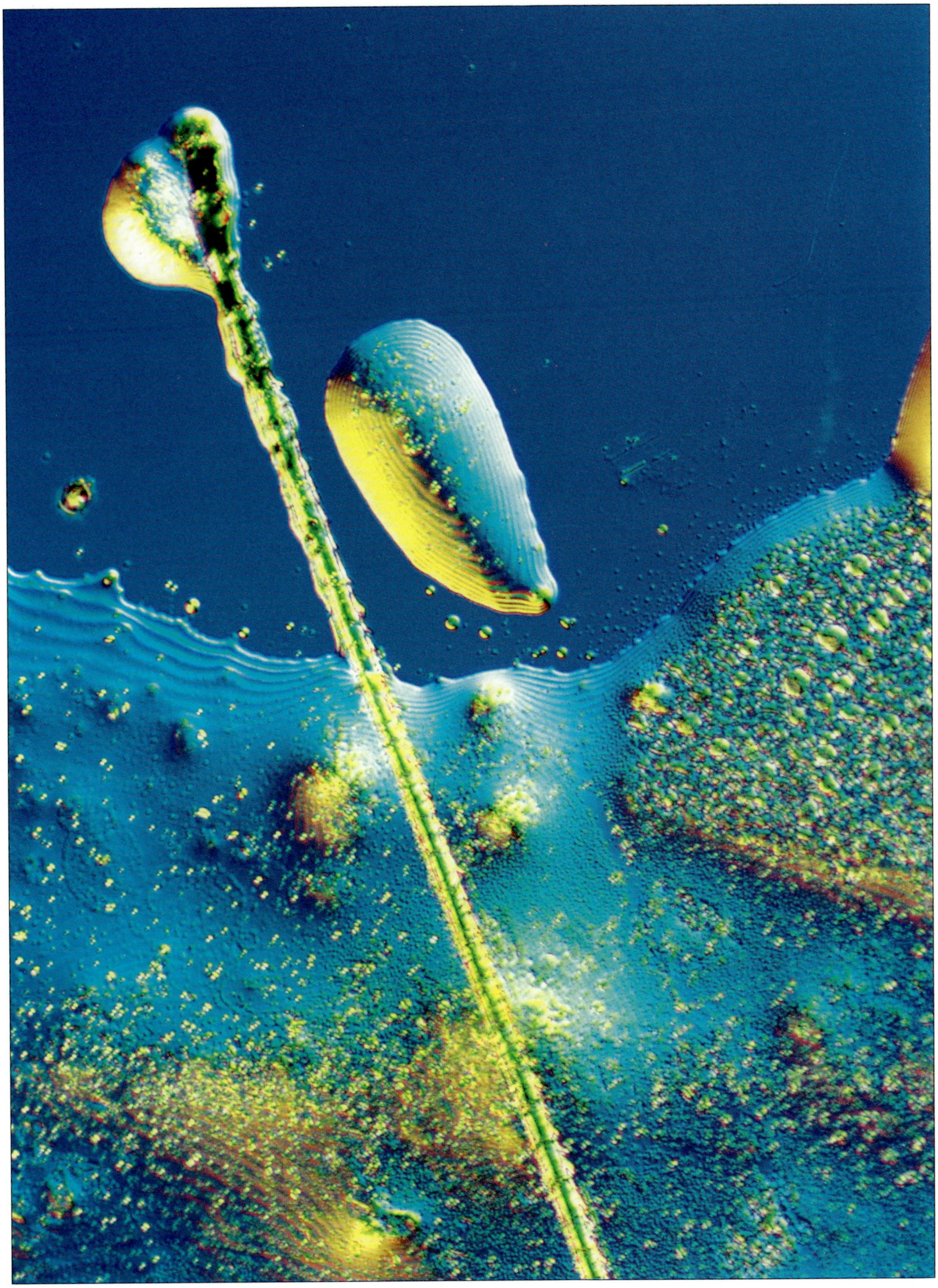

Buckyballs

Microprocessor 386

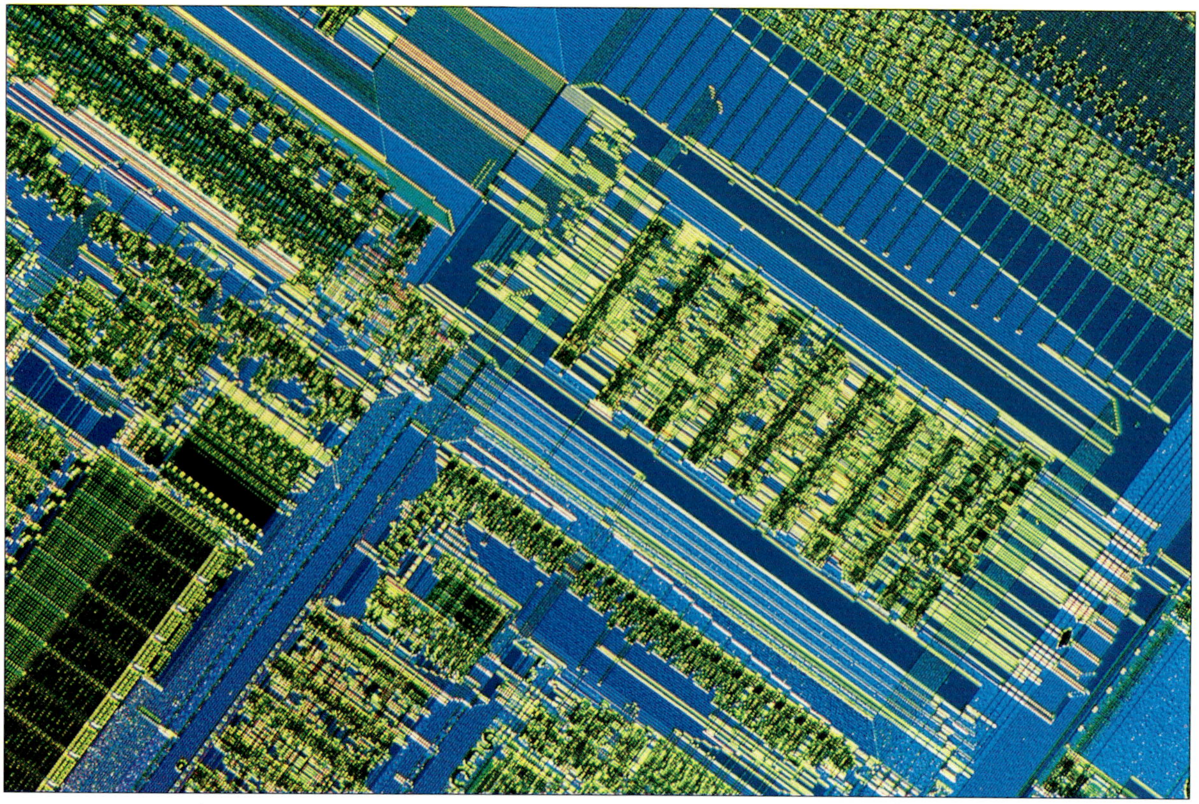

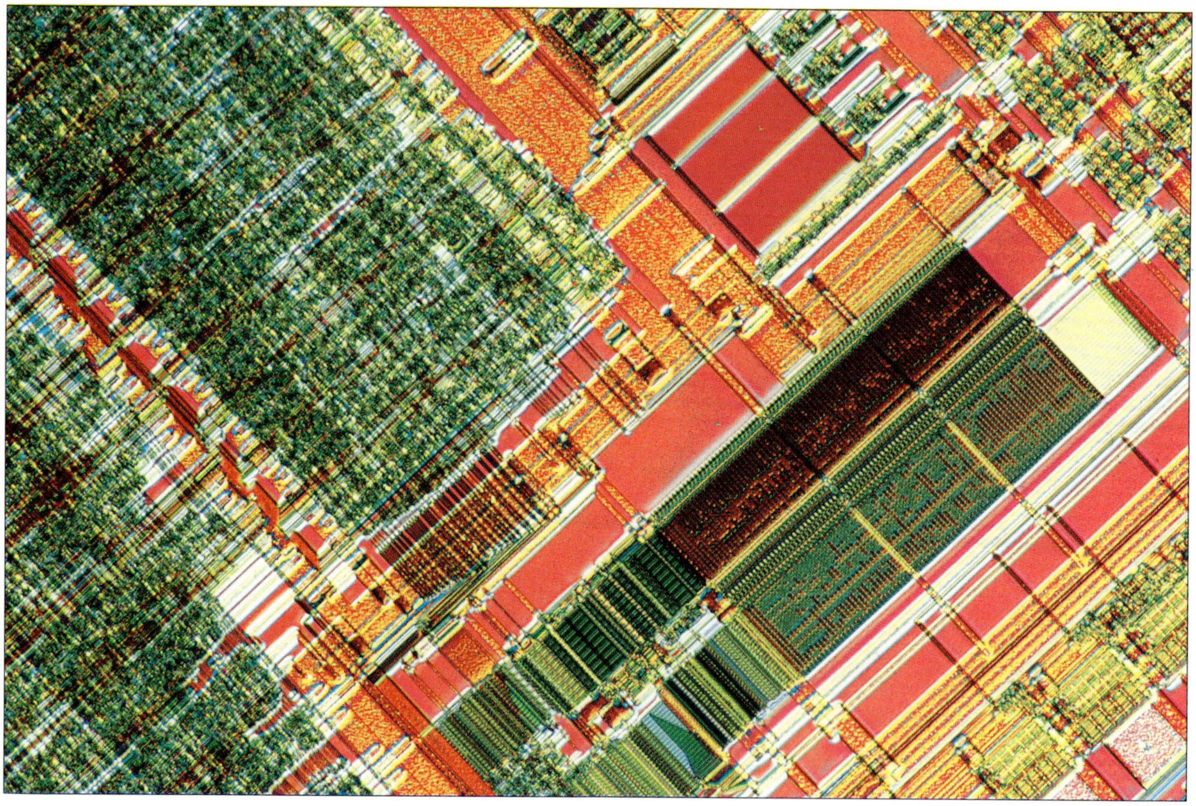

Microprocessor 486 SX

Microprocessor 486 DX

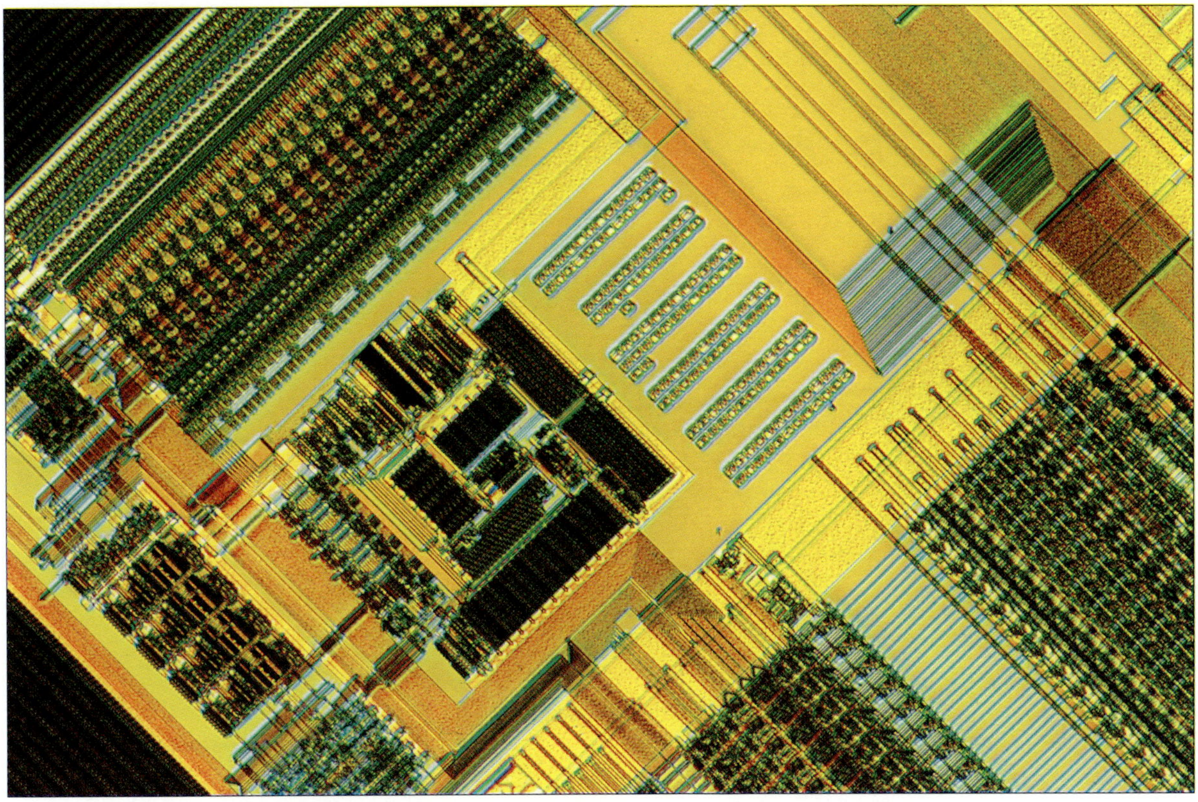

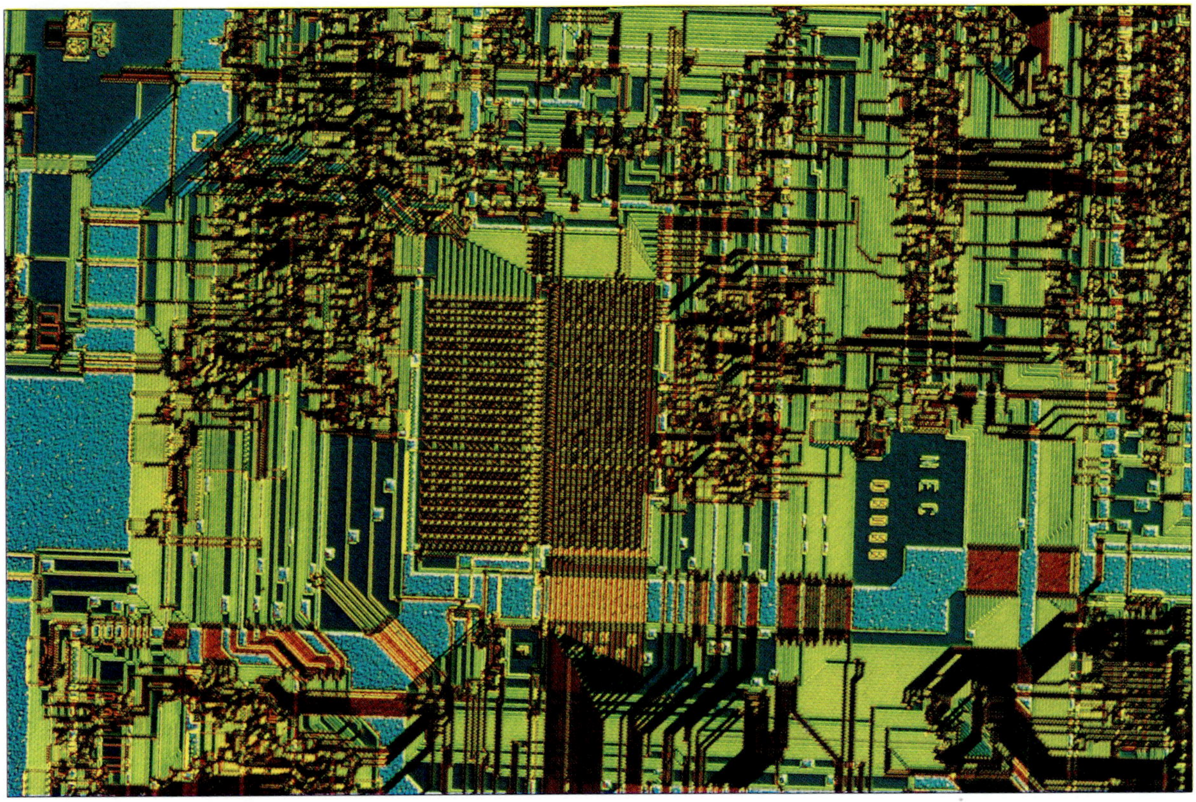

Microprocessor 8088

Lanthanum Aluminate

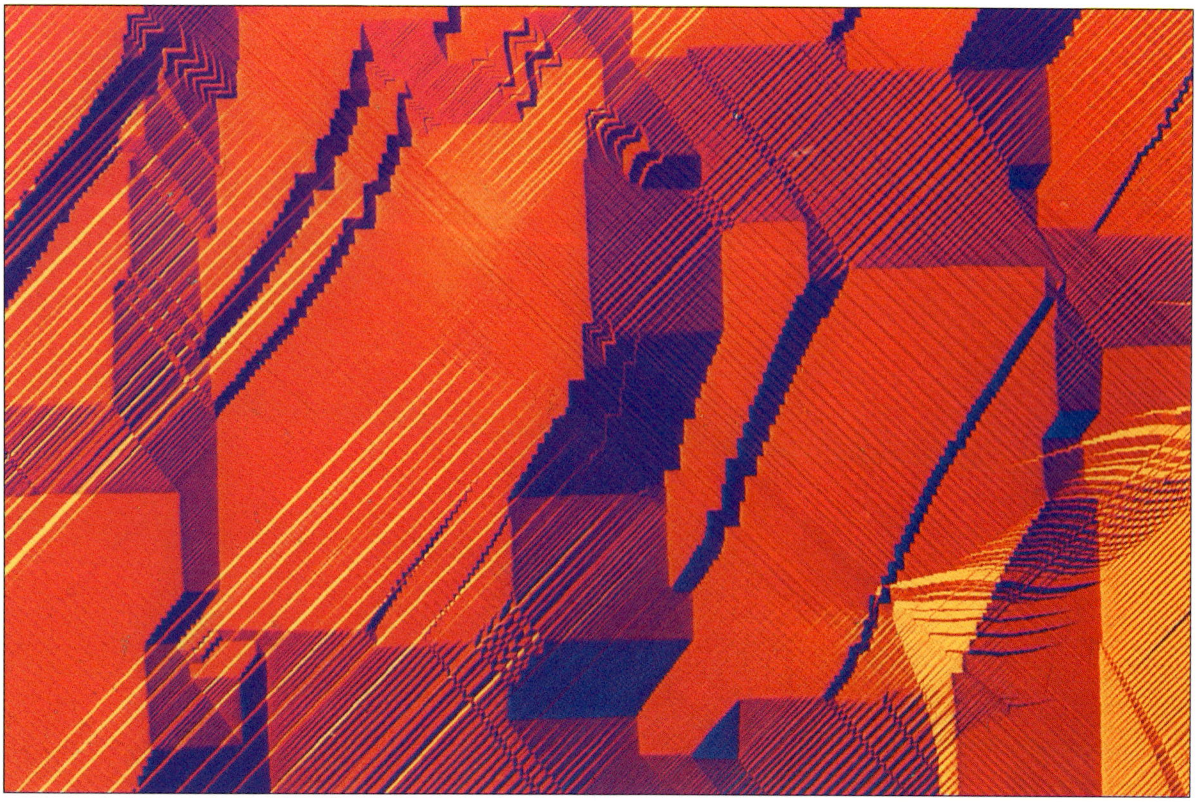

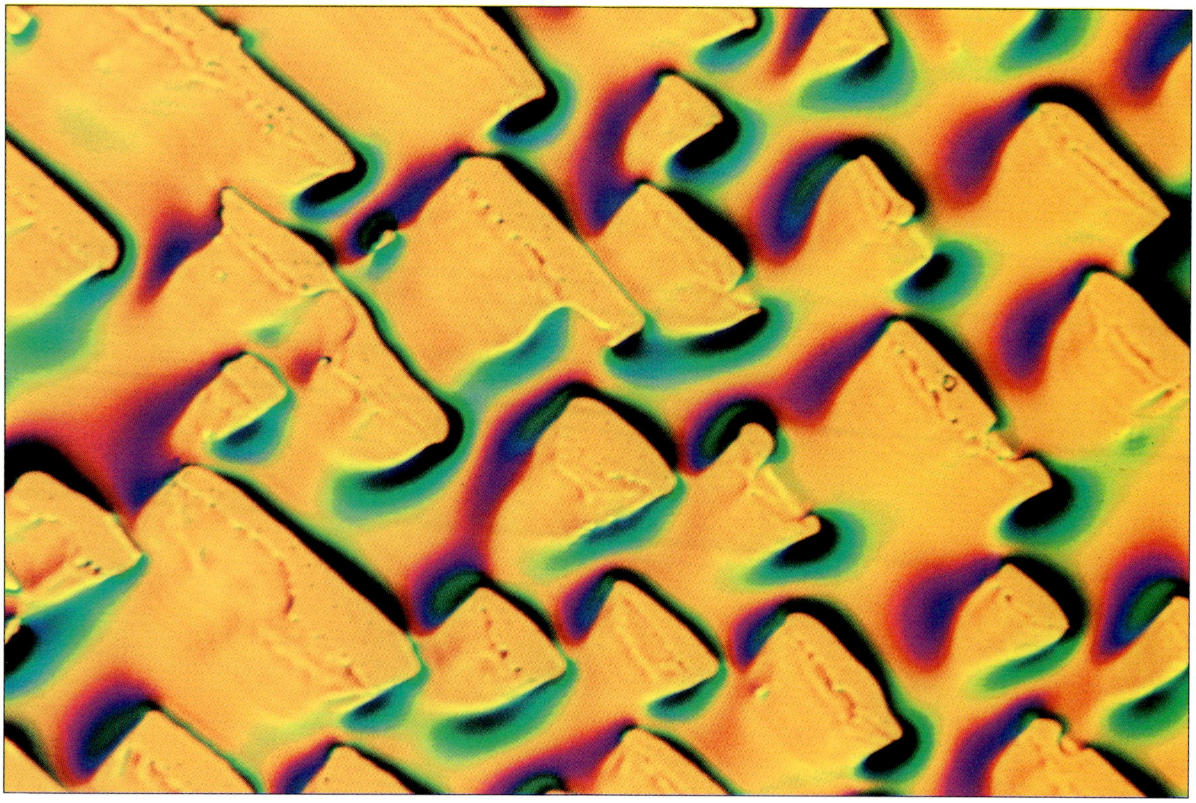

Lexan (Contact Lens Polymer)

Superconductor

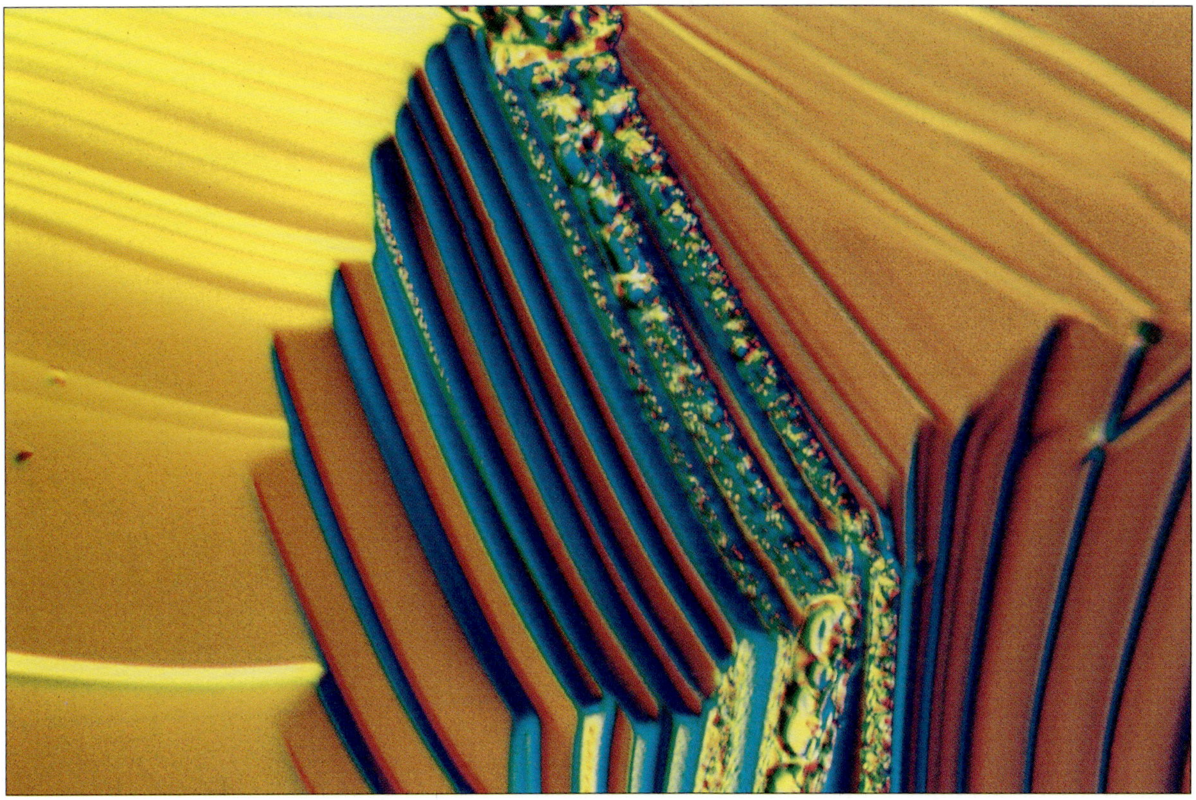

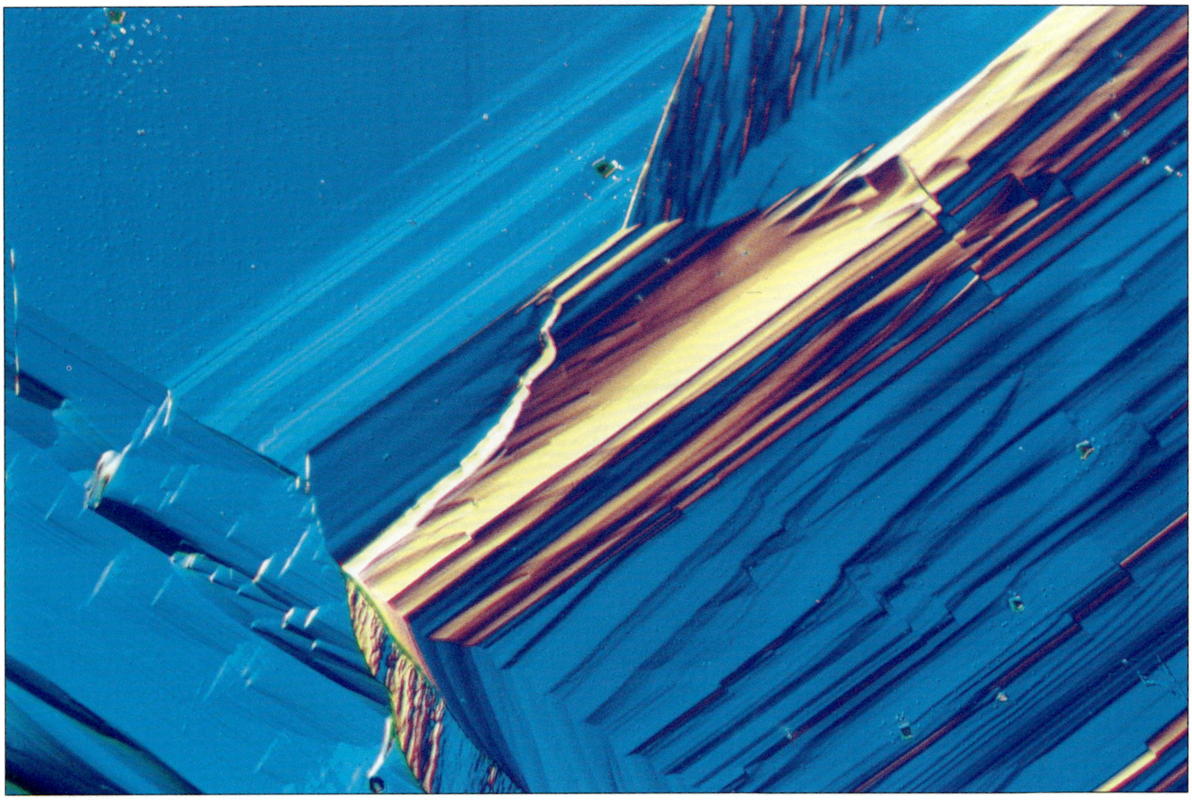

Organic Superconductor

Lanthanum Aluminate

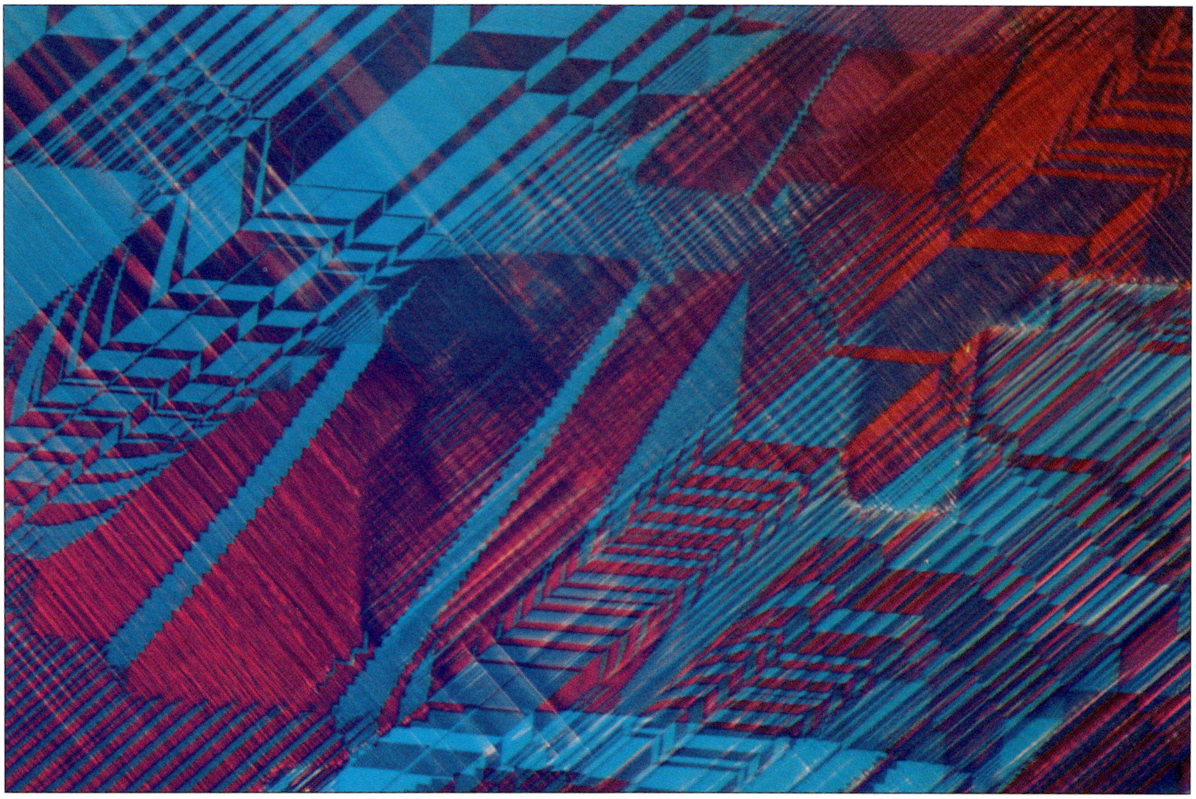

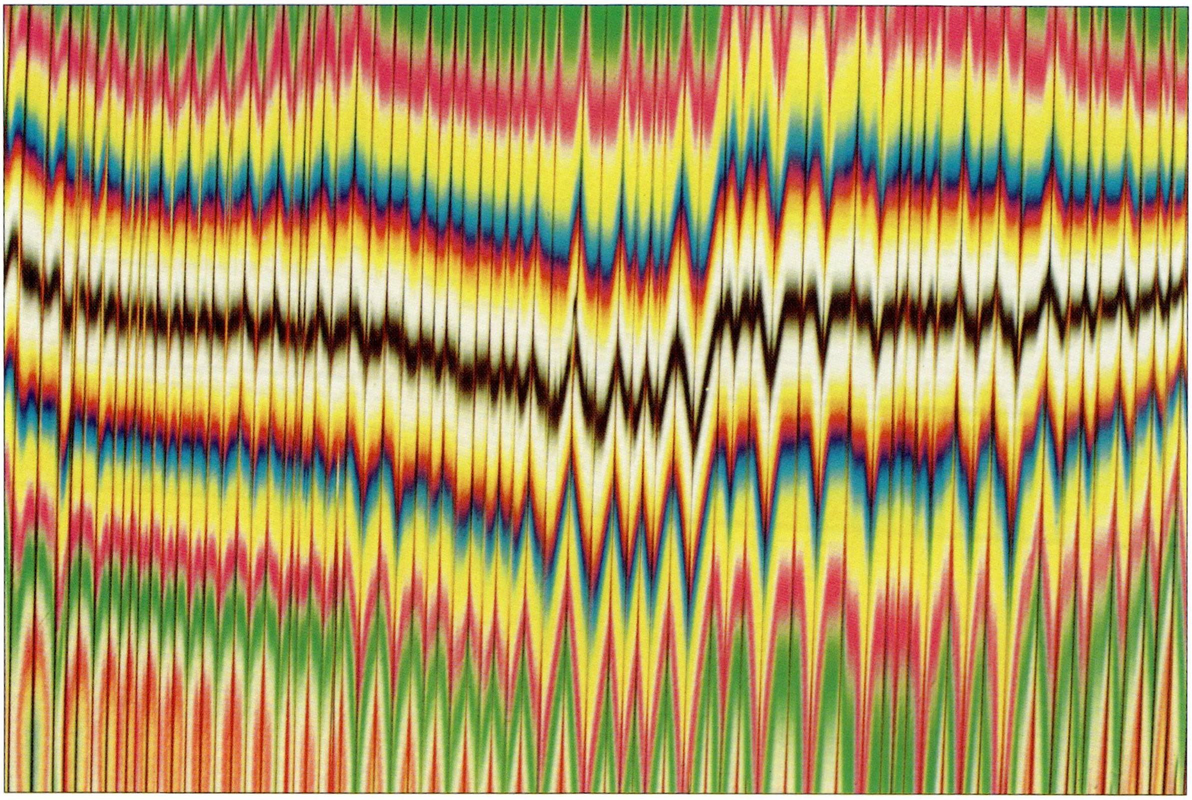

Lanthanum Aluminate

Lanthanum Aluminate

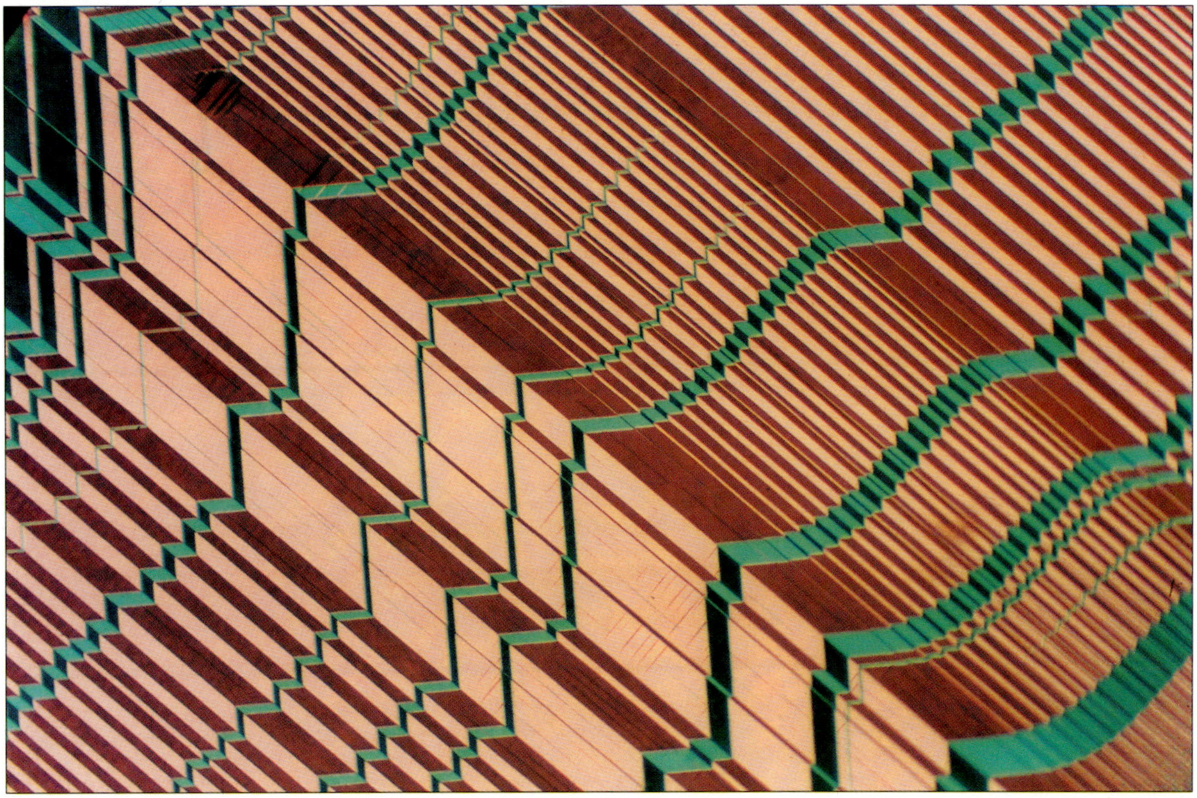

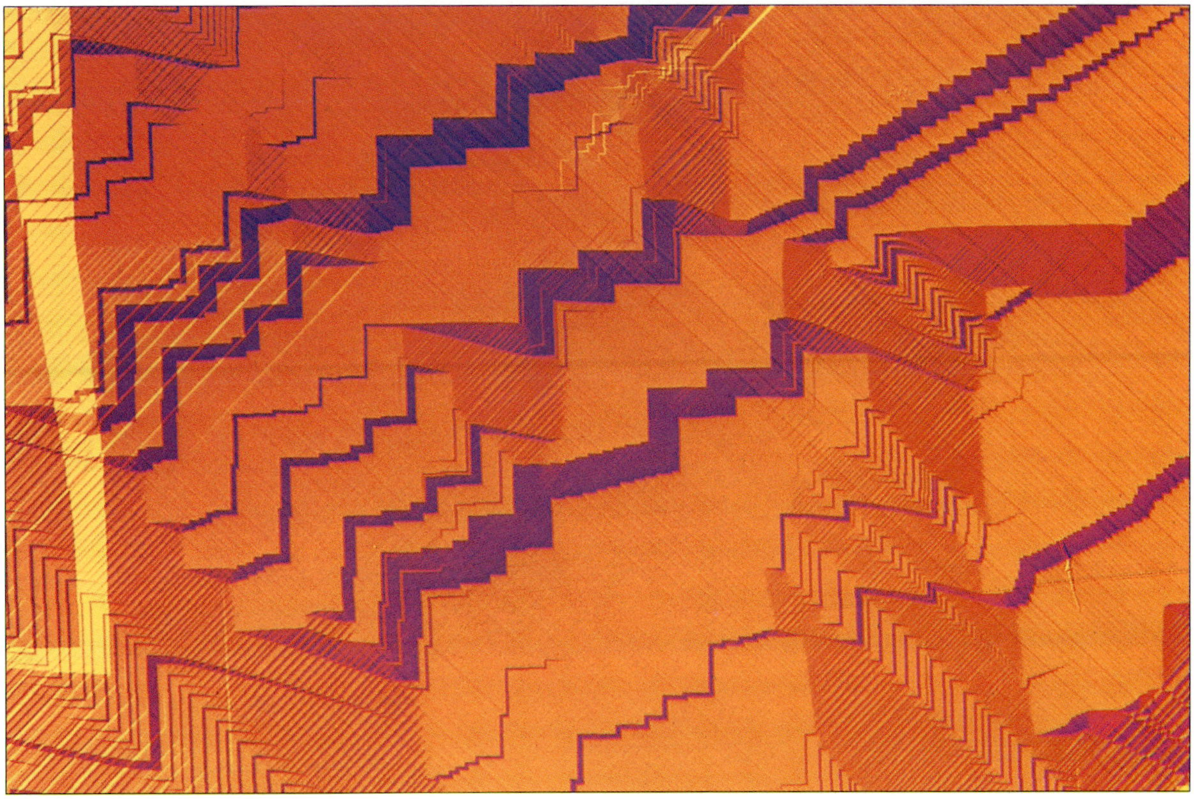

Lanthanum Aluminate

Table Salt

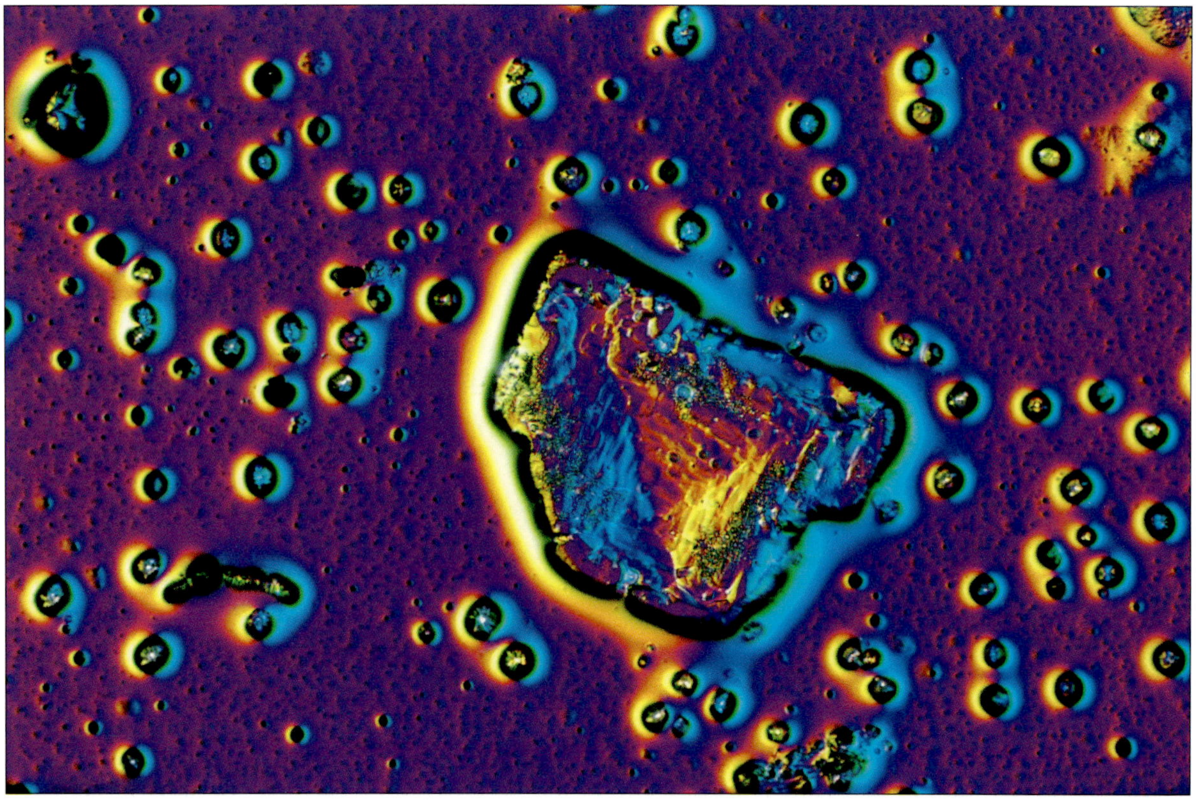

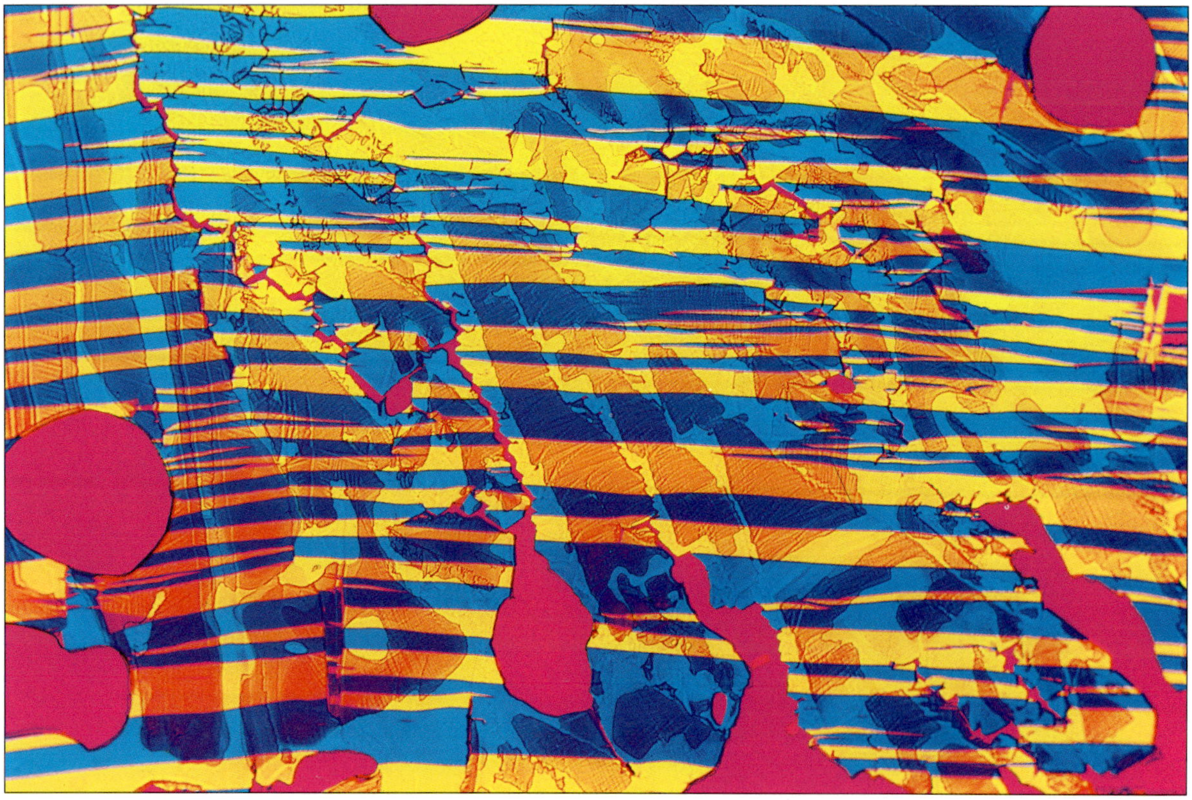

Sugar

Buckyballs

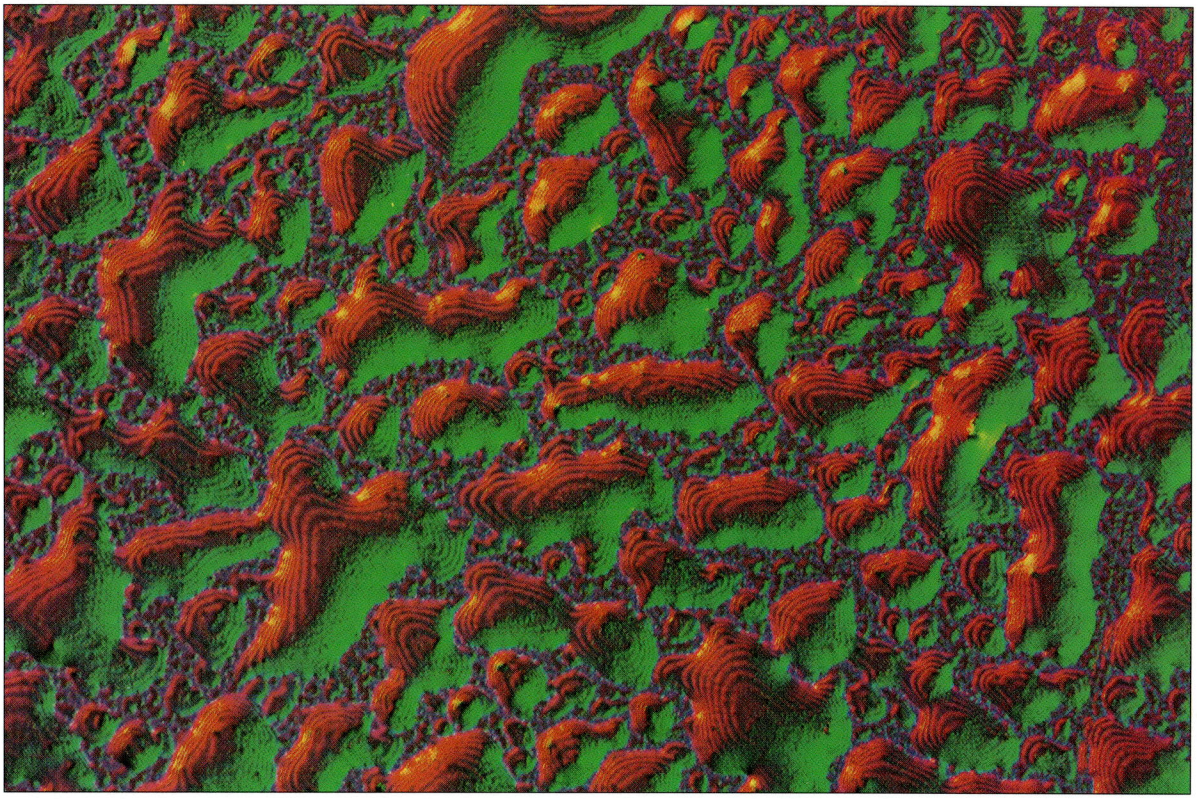

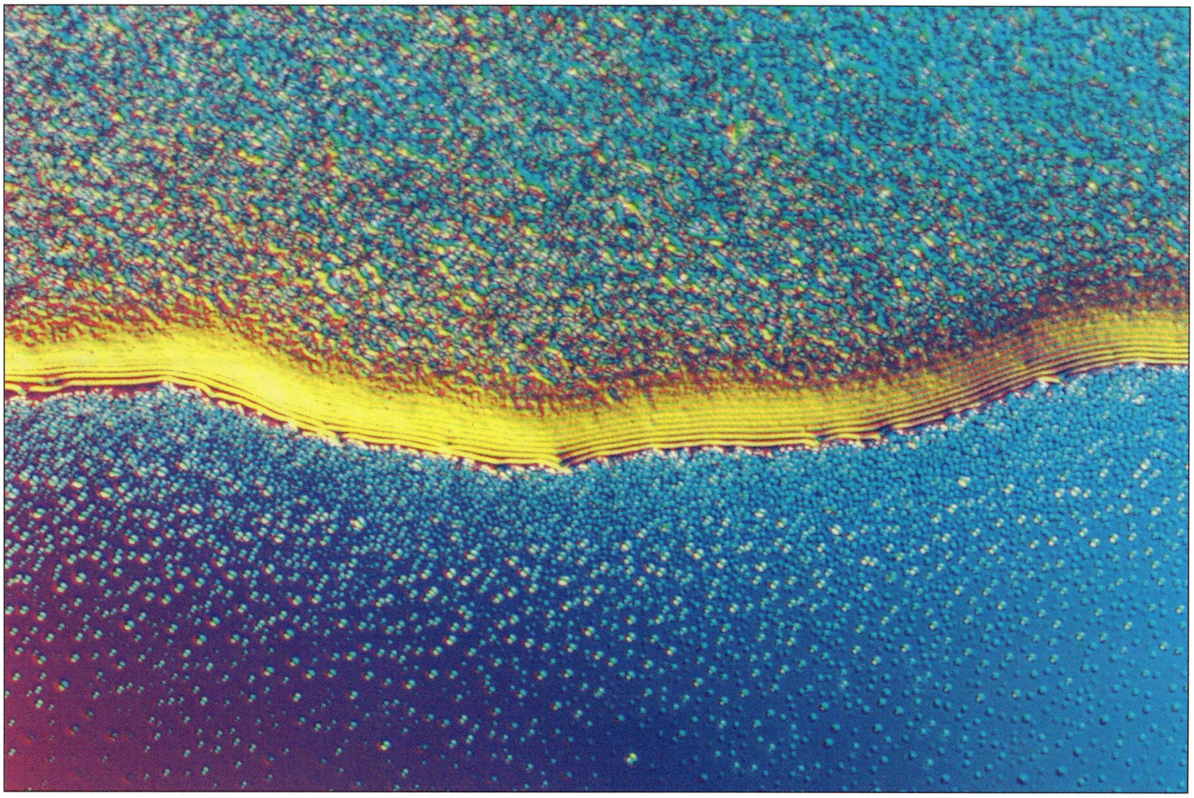

Buckyballs

Dye

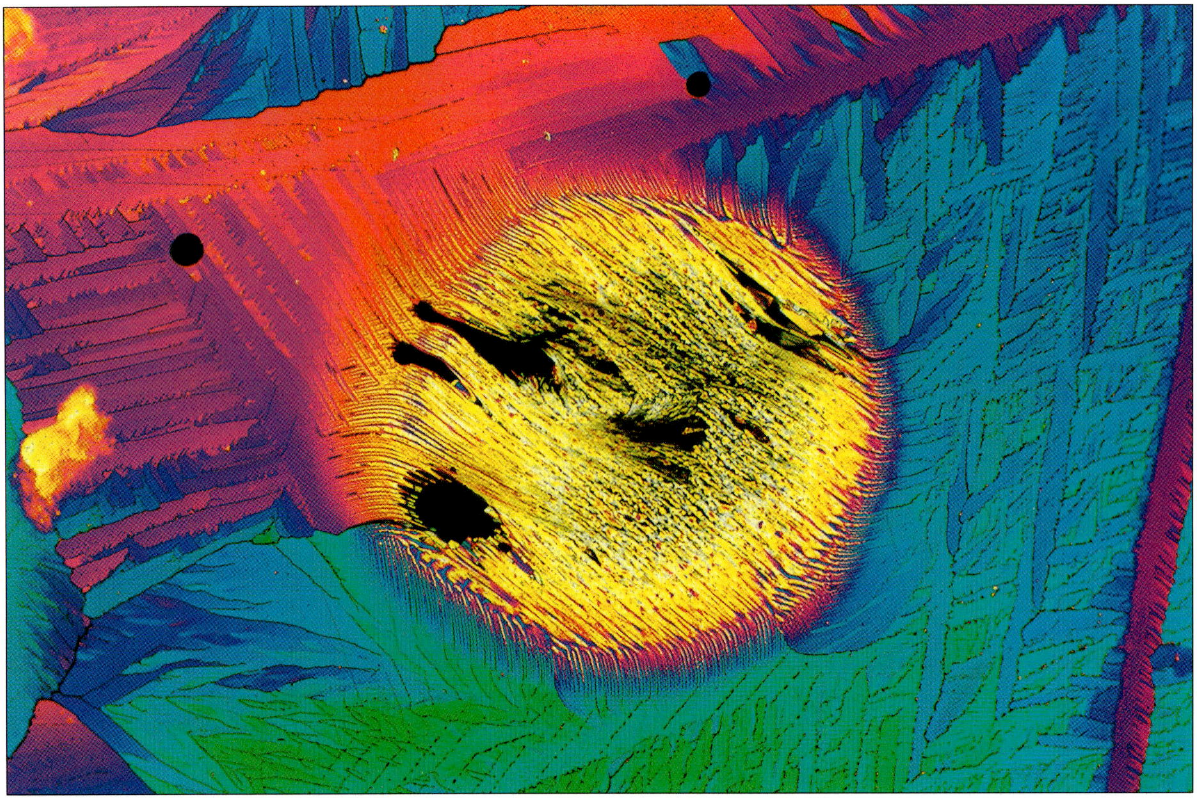

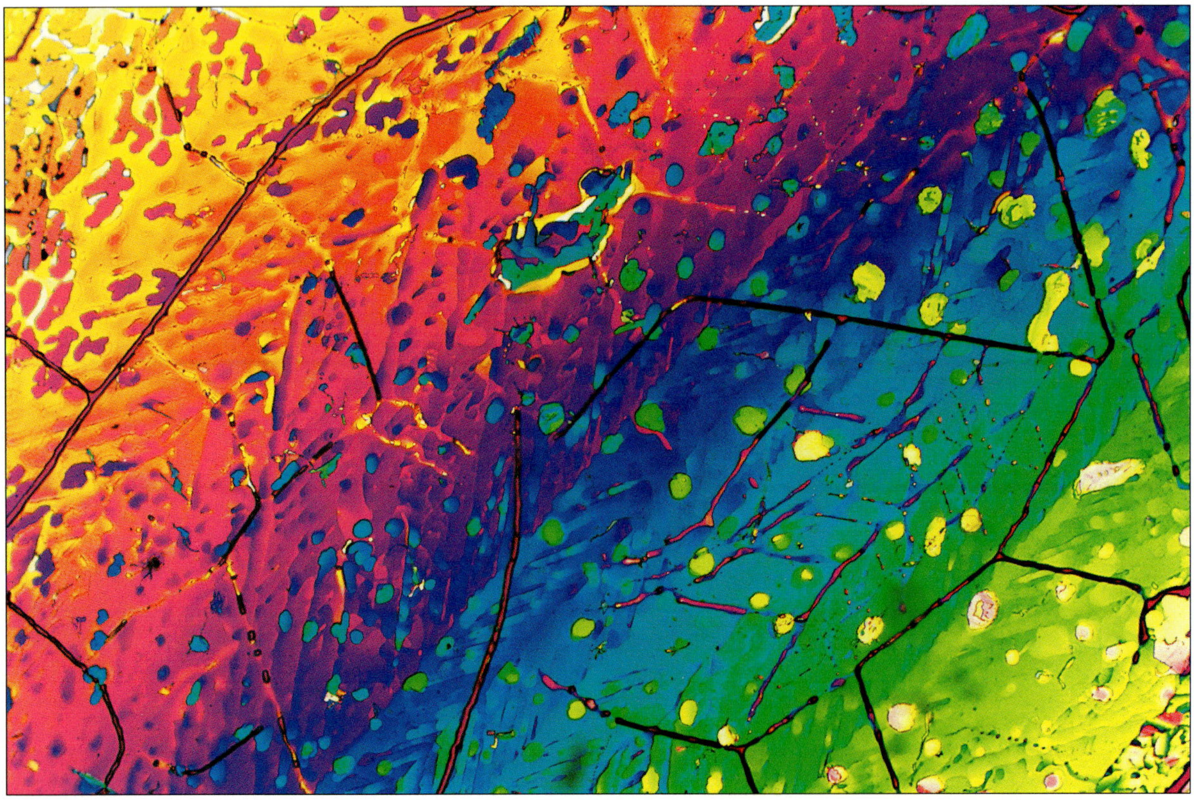

Dye

Cinnamon

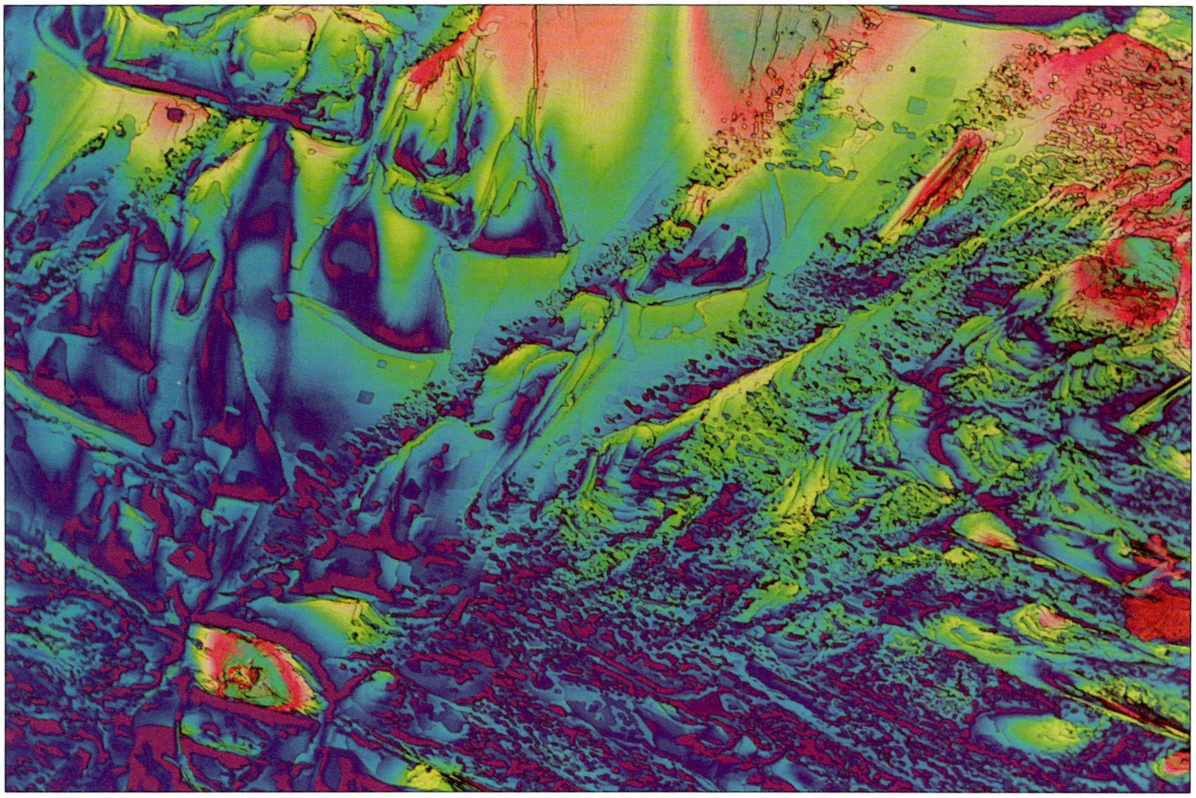

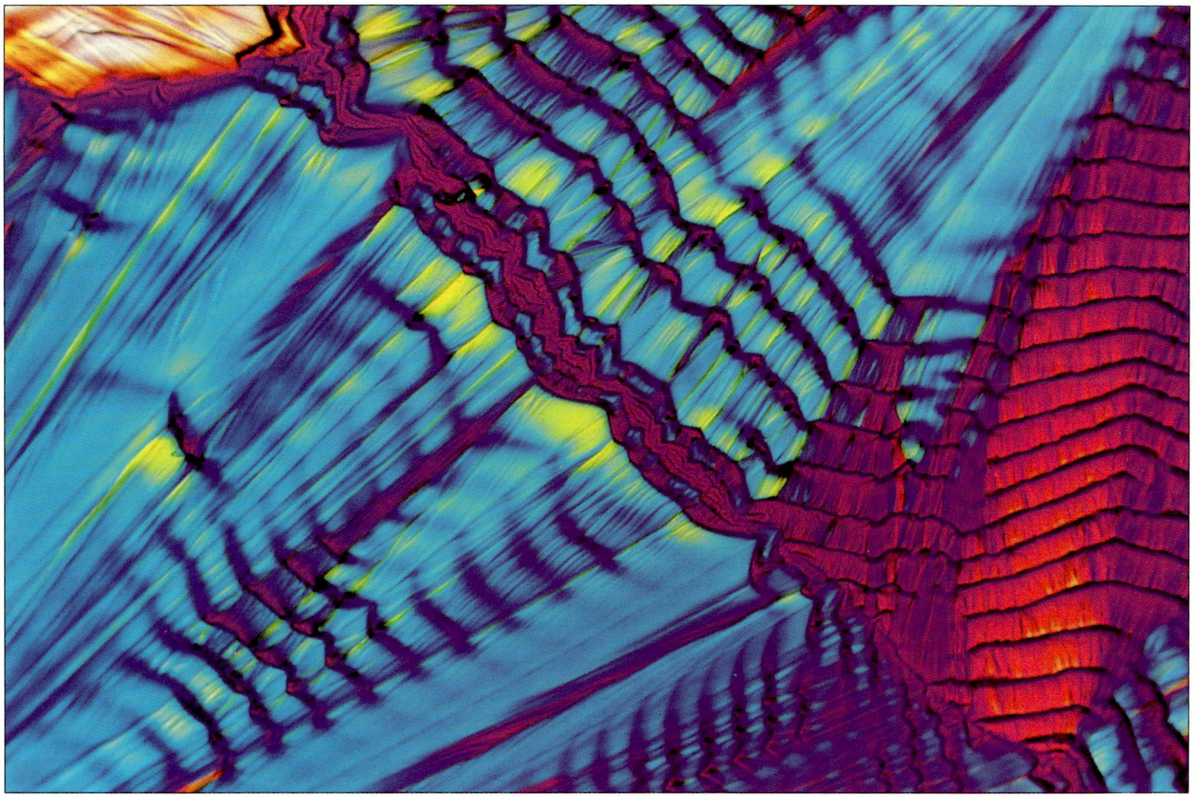

Malt

Orange

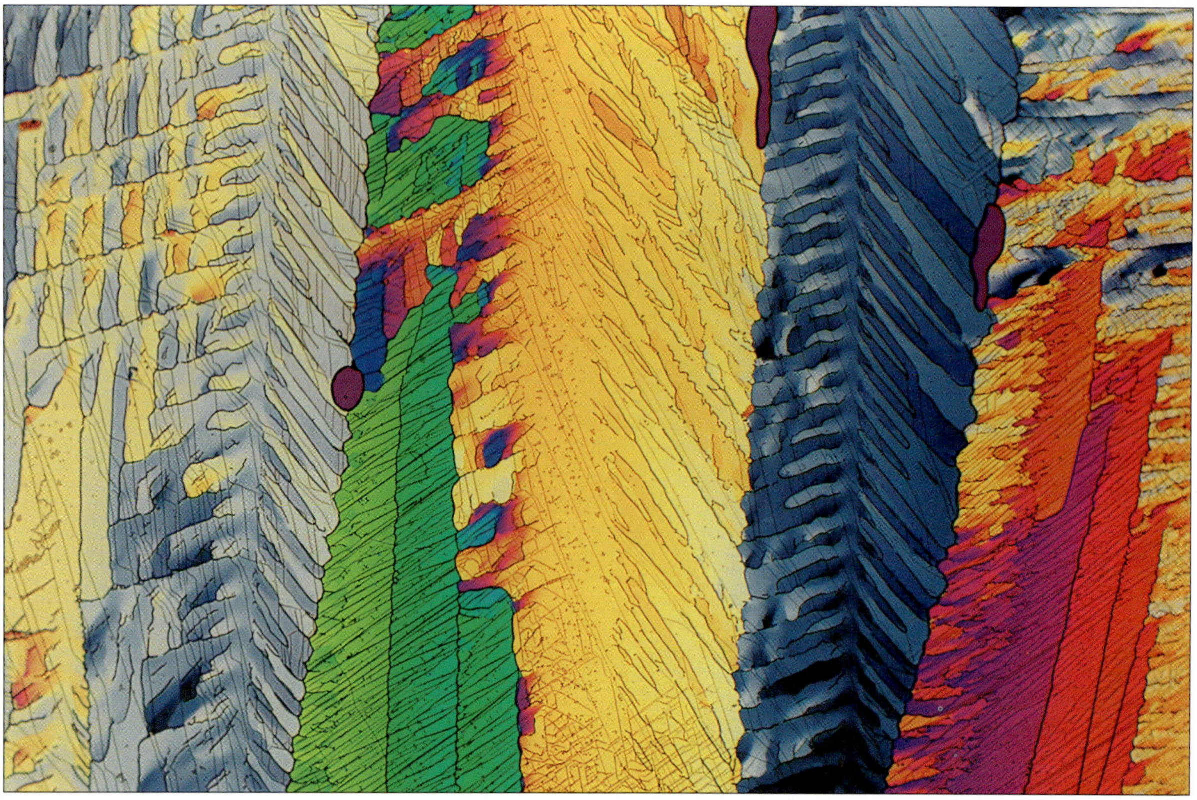

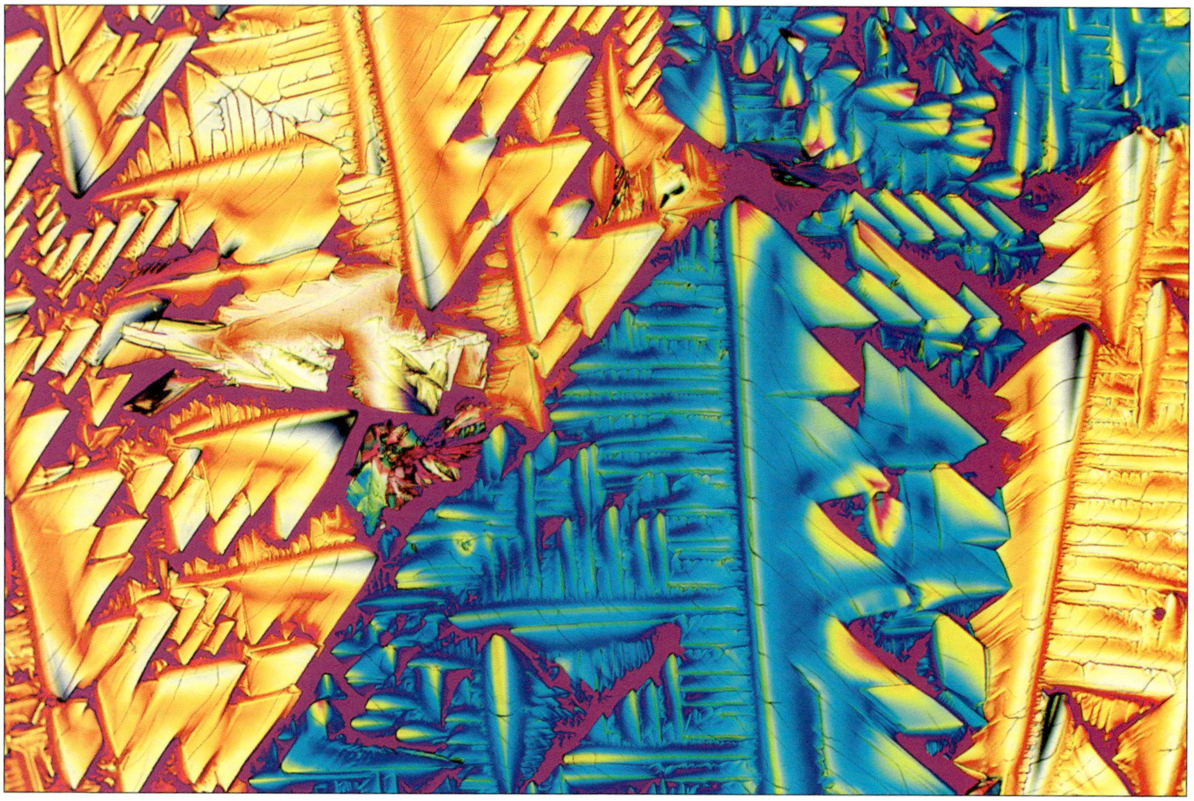

Apple

Cherry

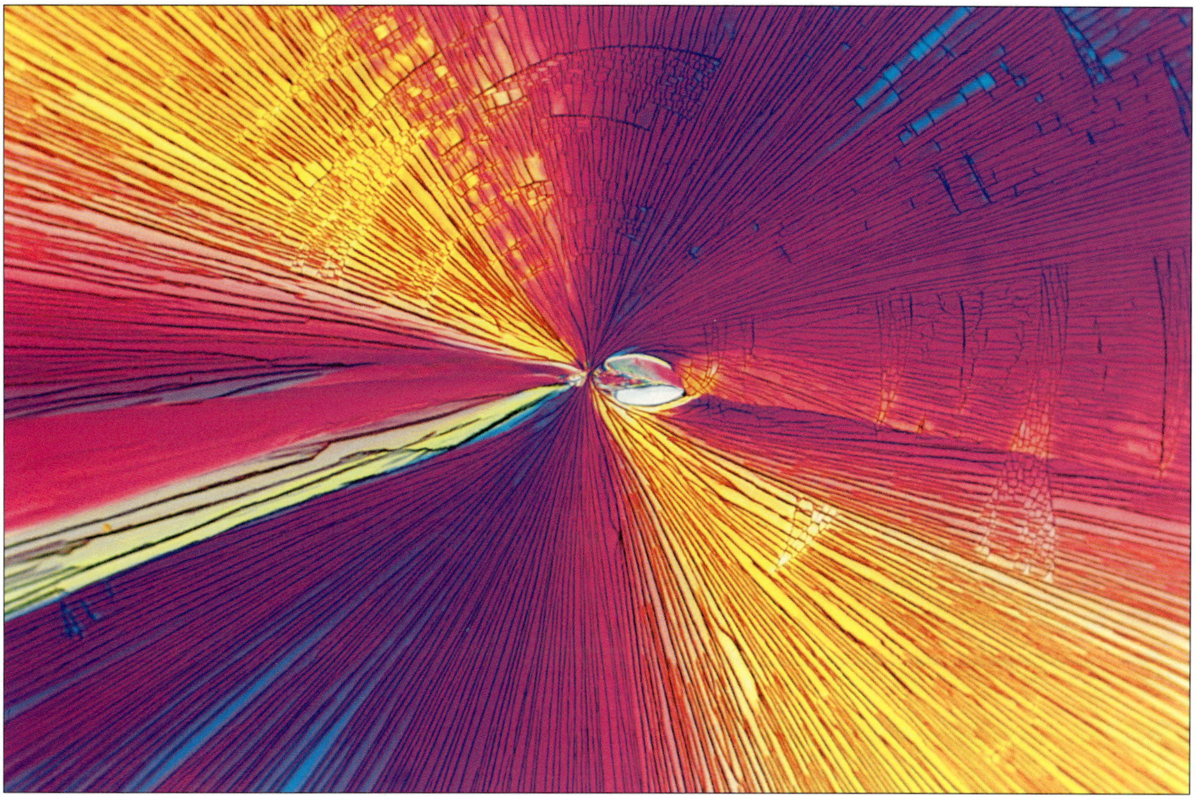

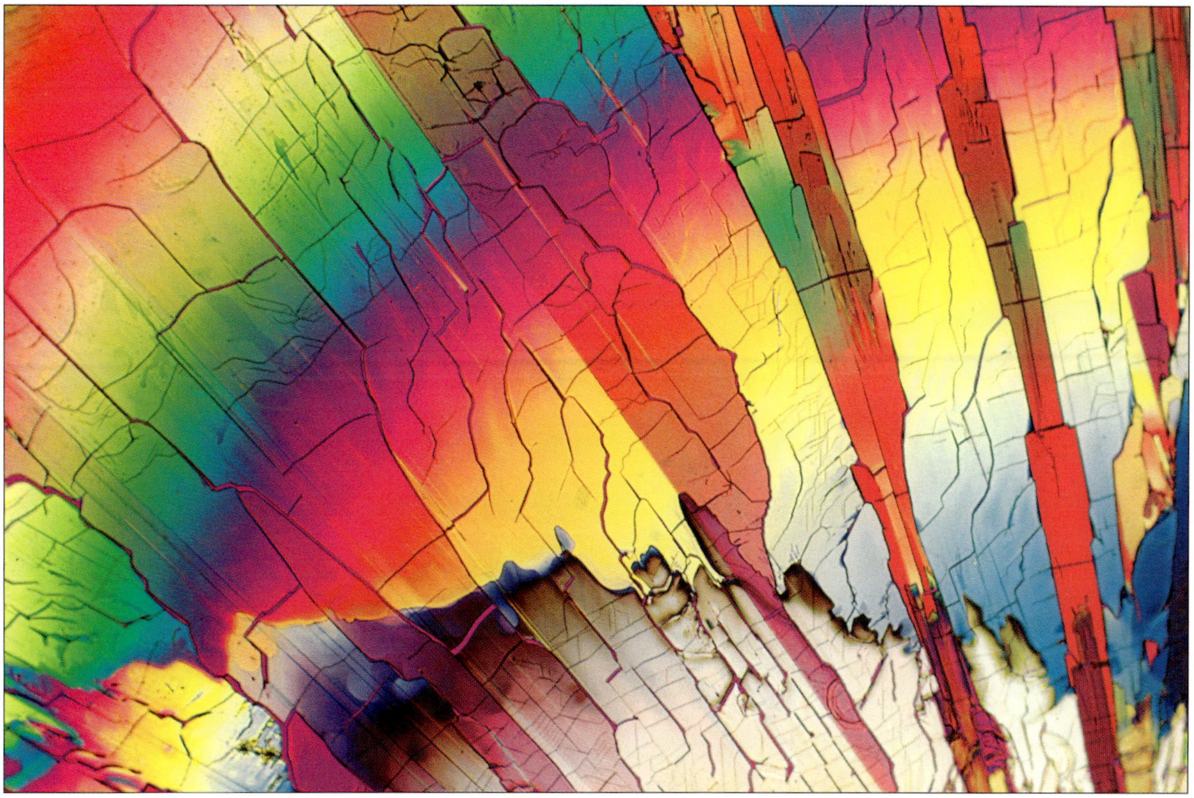

Pineapple

Thiamine (Vitamin B1)

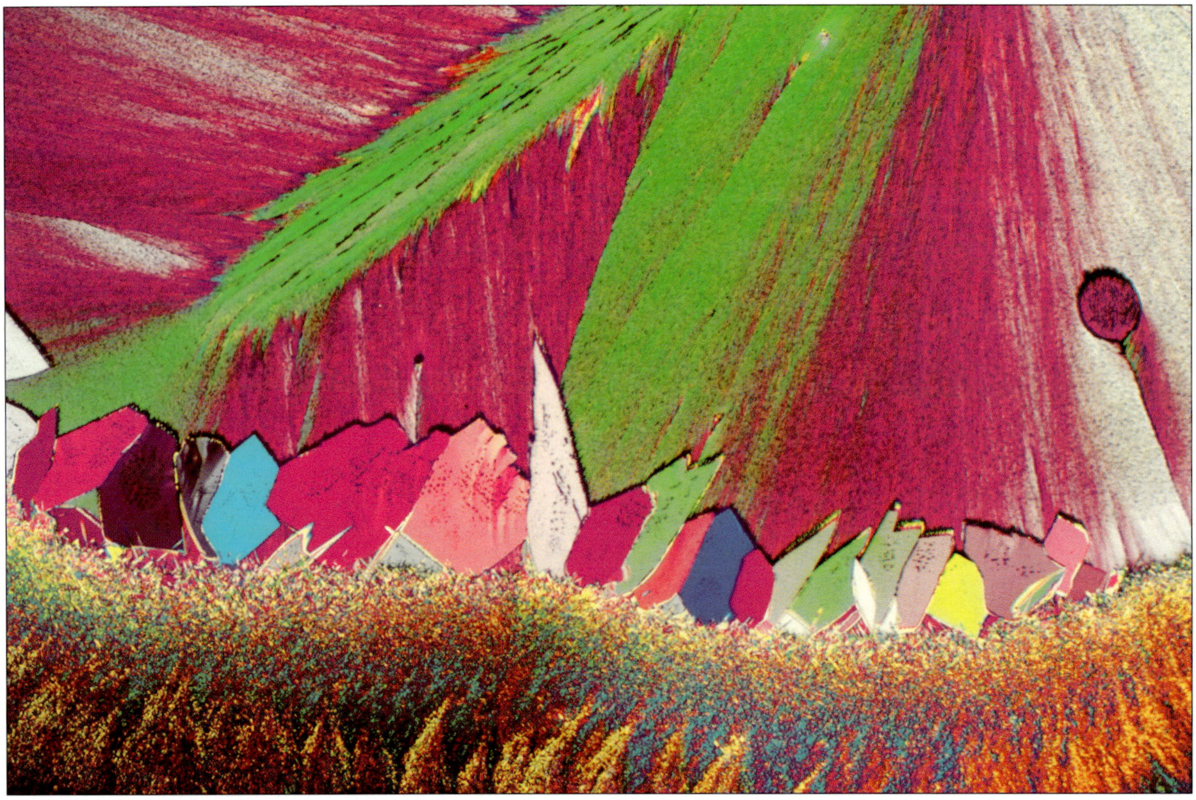

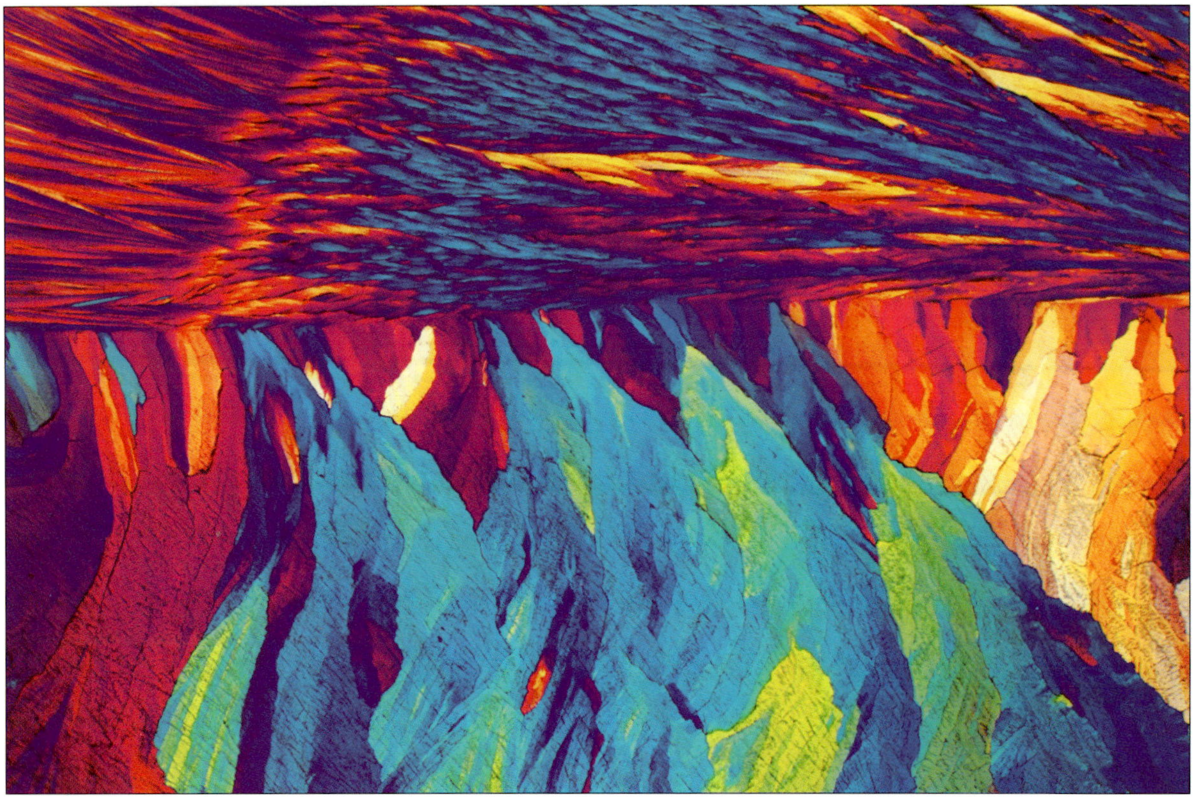

Cholecalciferol (Vitamin D3)

Alpha Tocopherol (Vitamin E)

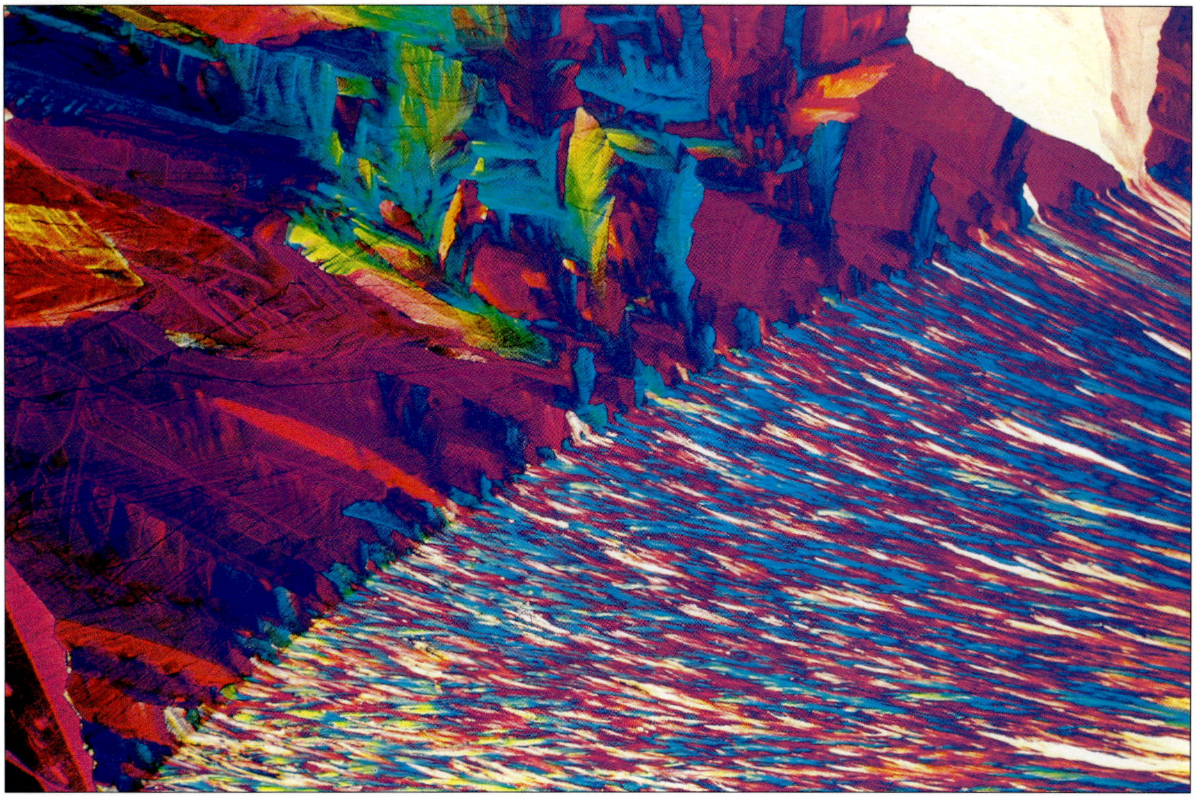

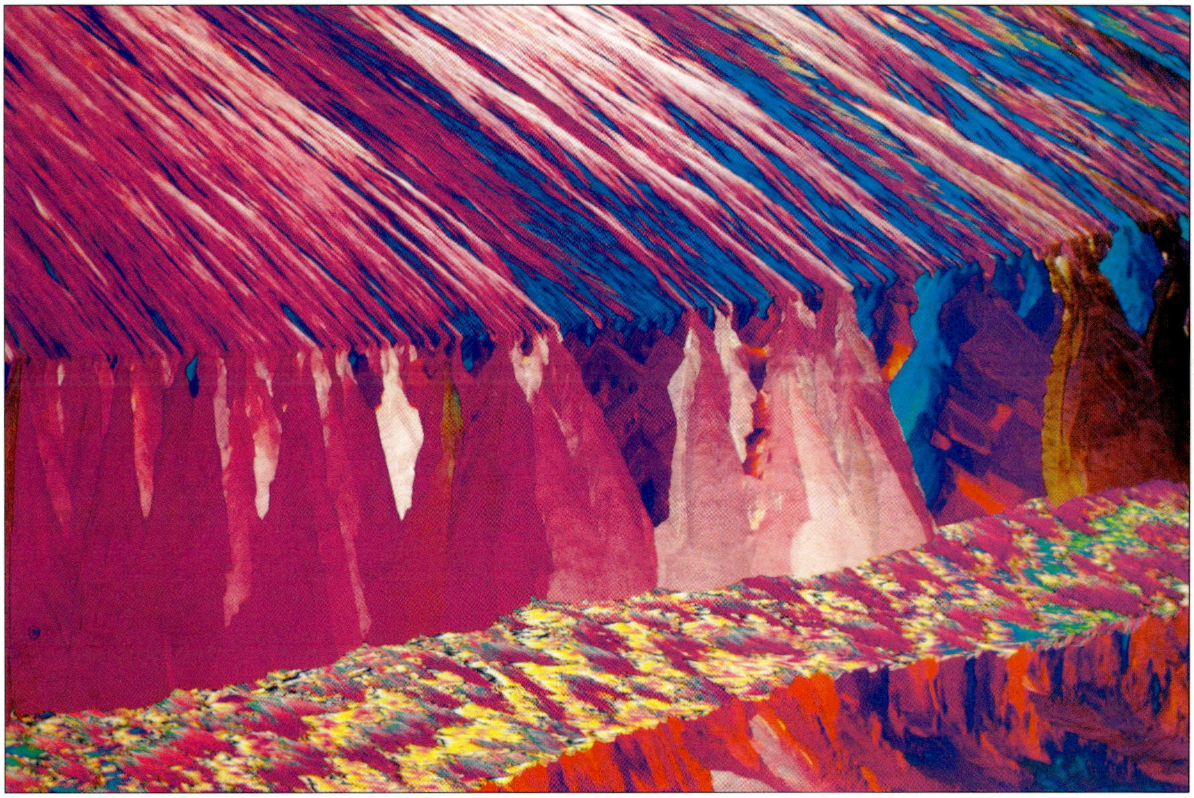

Alpha Tocopherol (Vitamin E)

Cocaine HCl

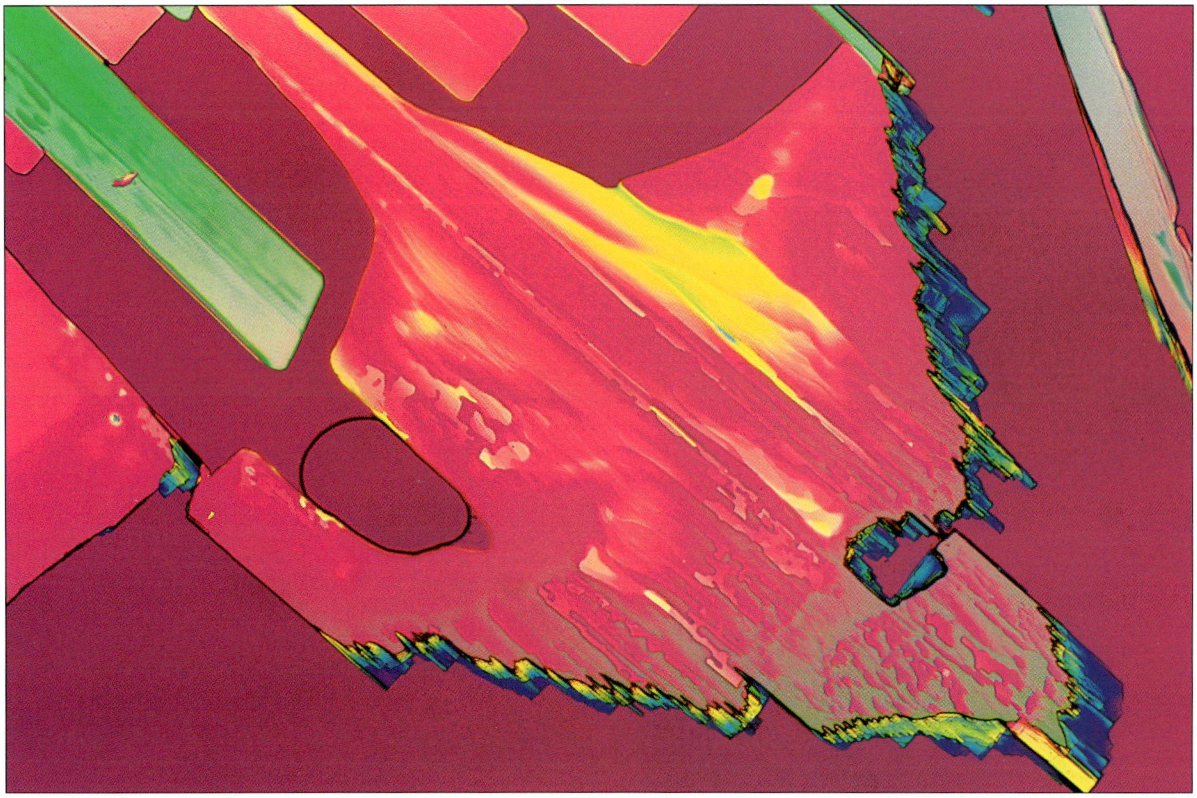

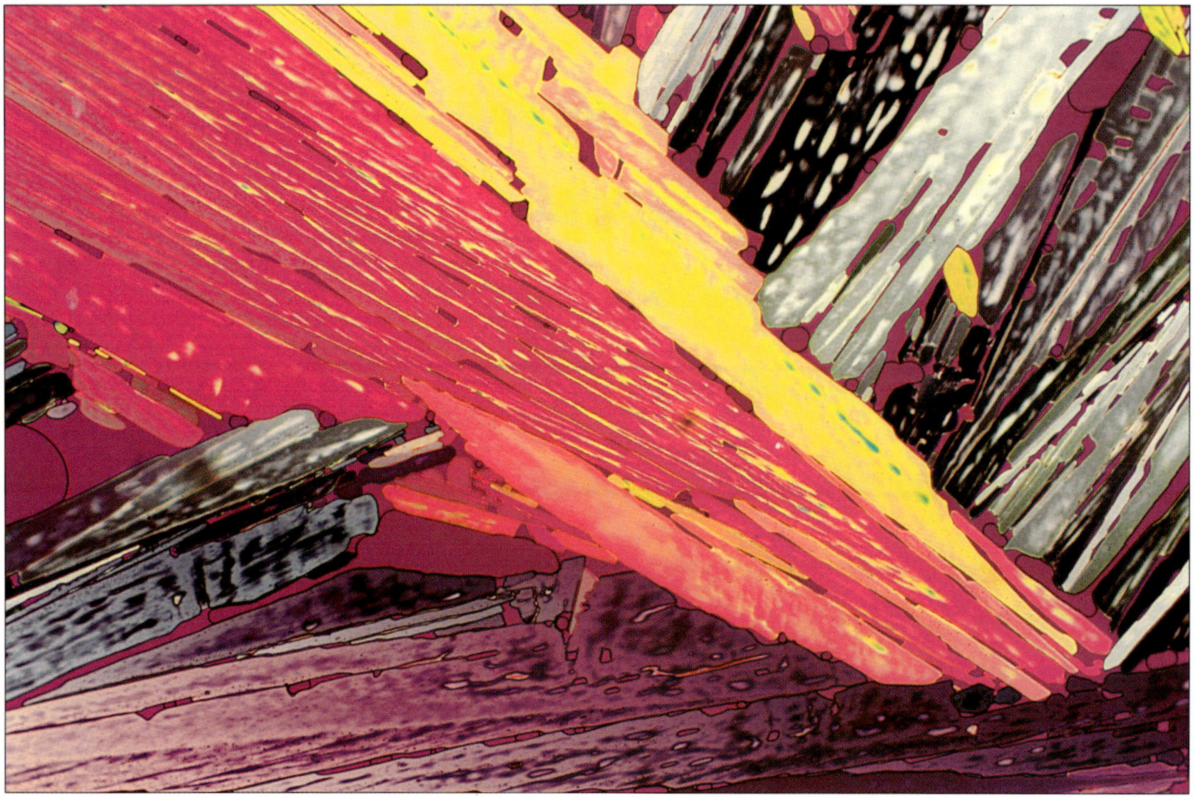

Cocaine HCl

Alpha Tocopherol (Vitamin E)

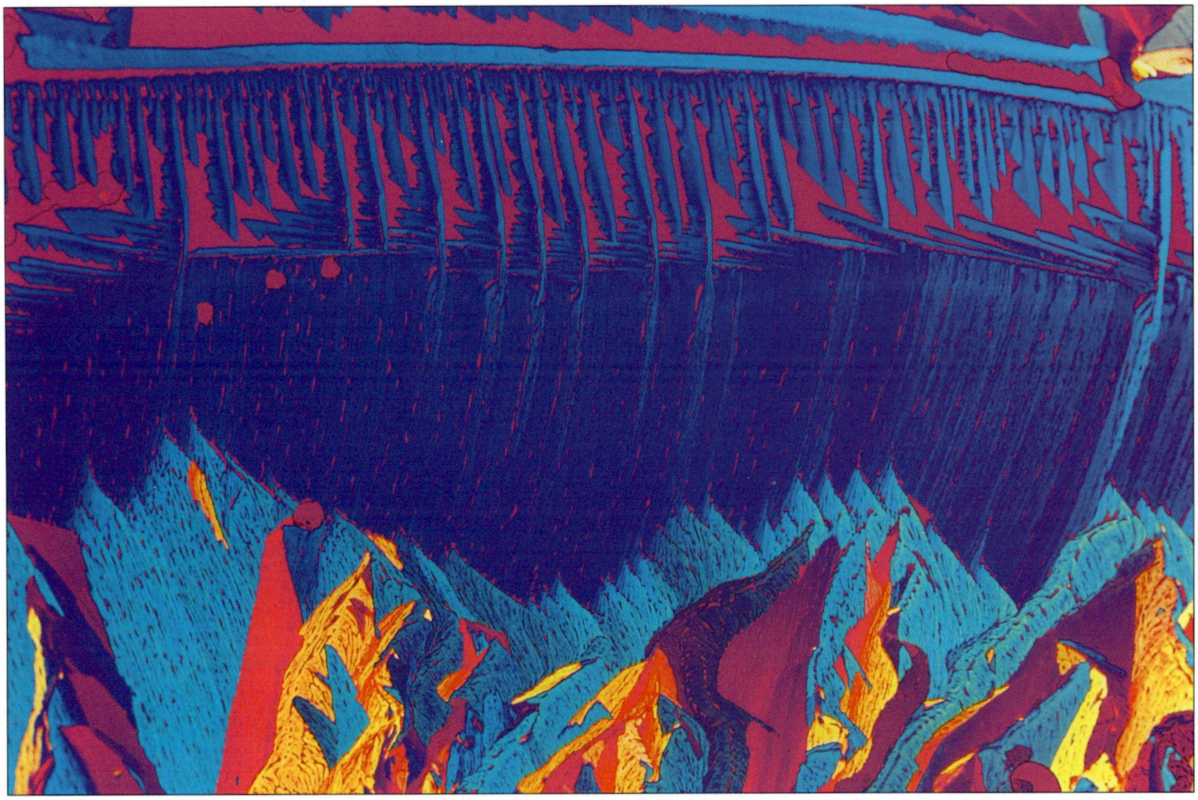

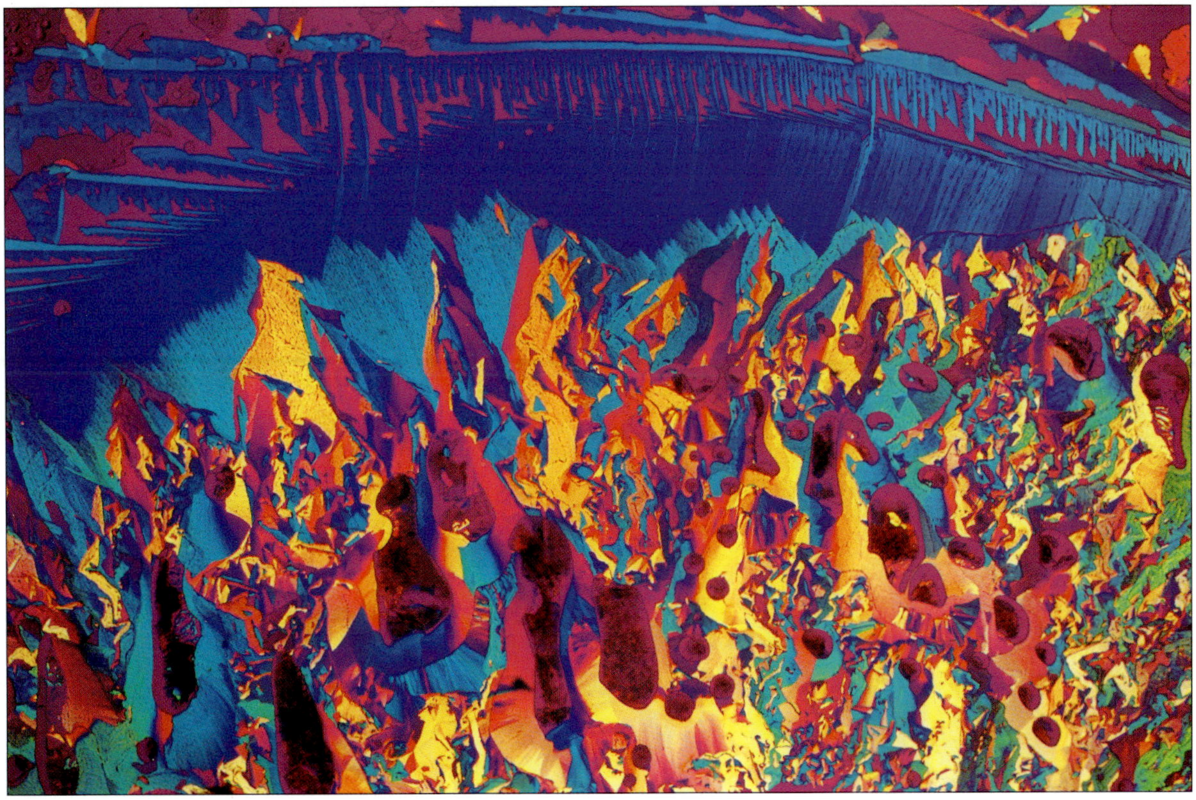

Alpha Tocopherol (Vitamin E)

Alpha Tocopherol (Vitamin E)

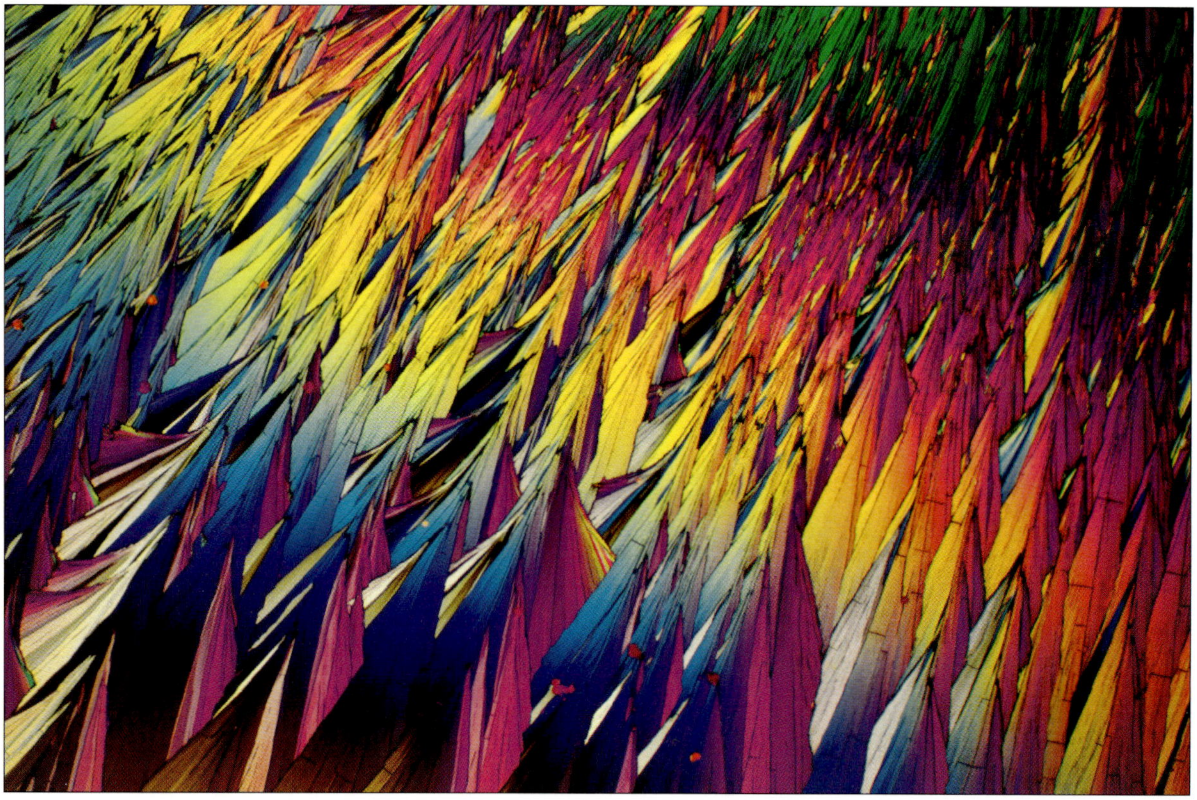

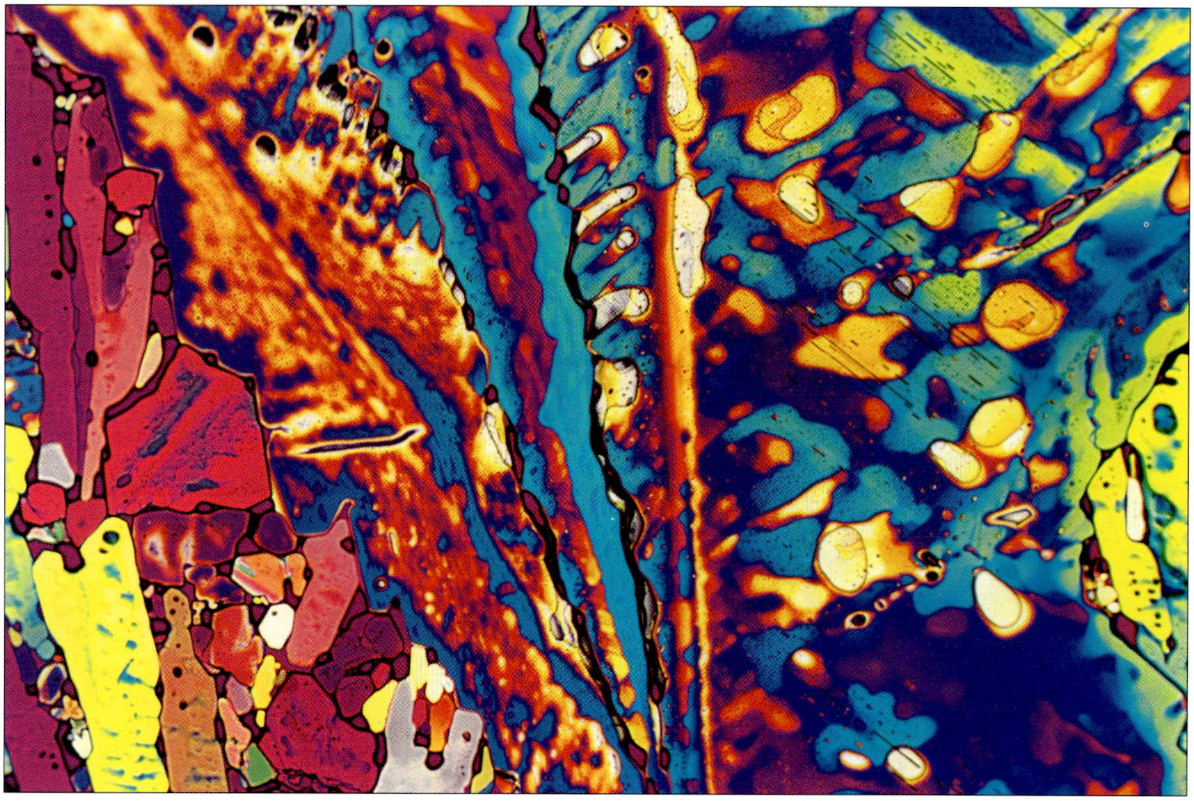

Cyanocobalamin

Silicon

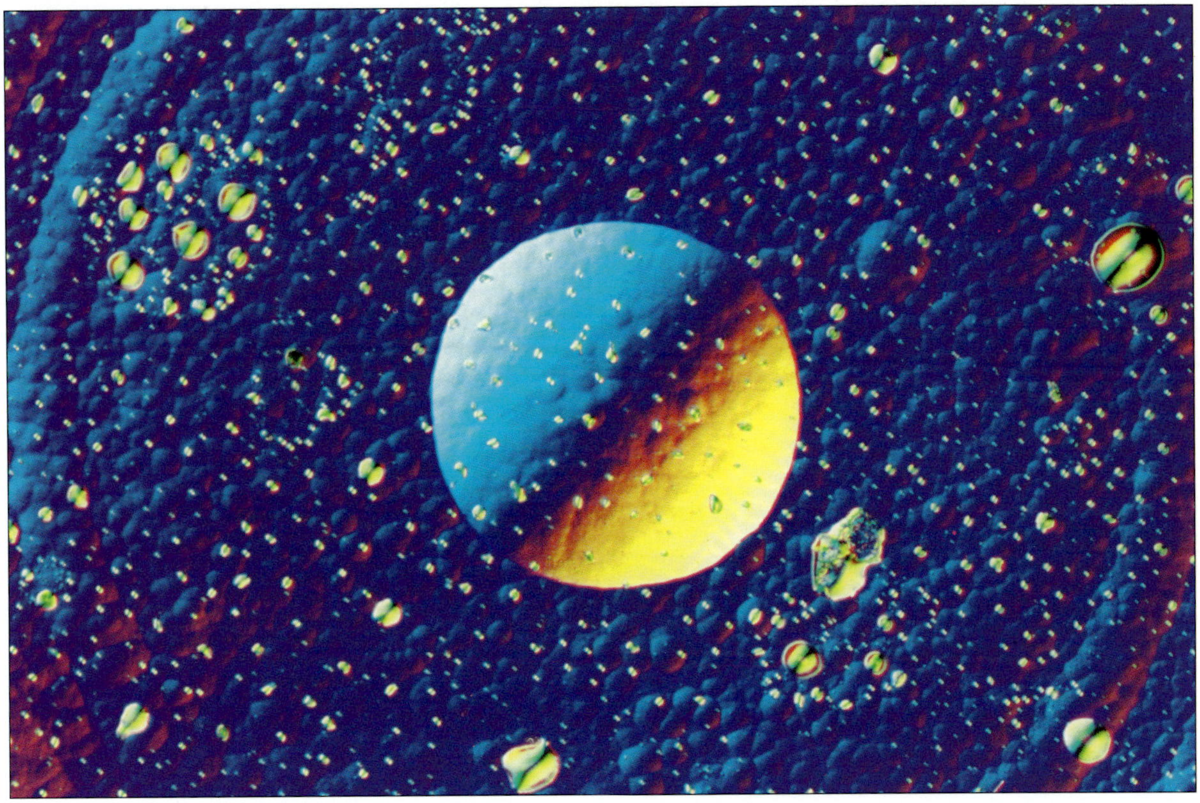

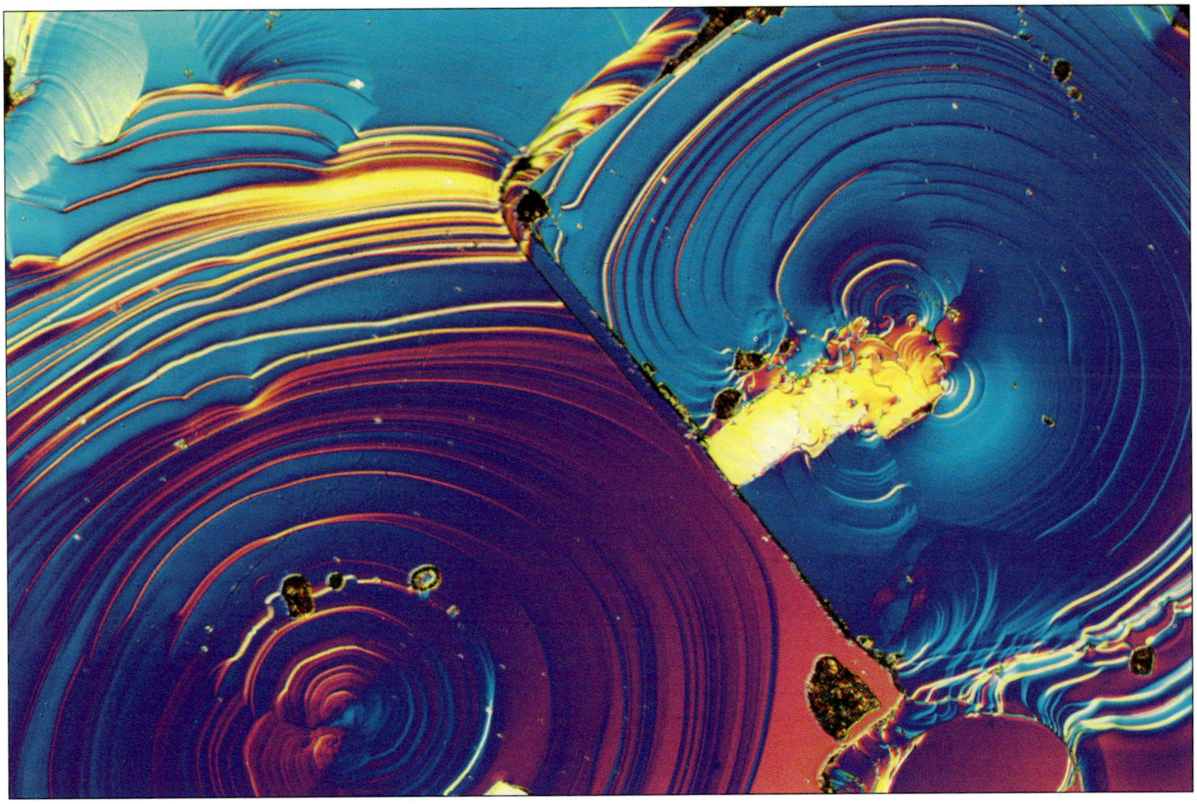

Gadolinium

Riboflavin (Vitamin B2)

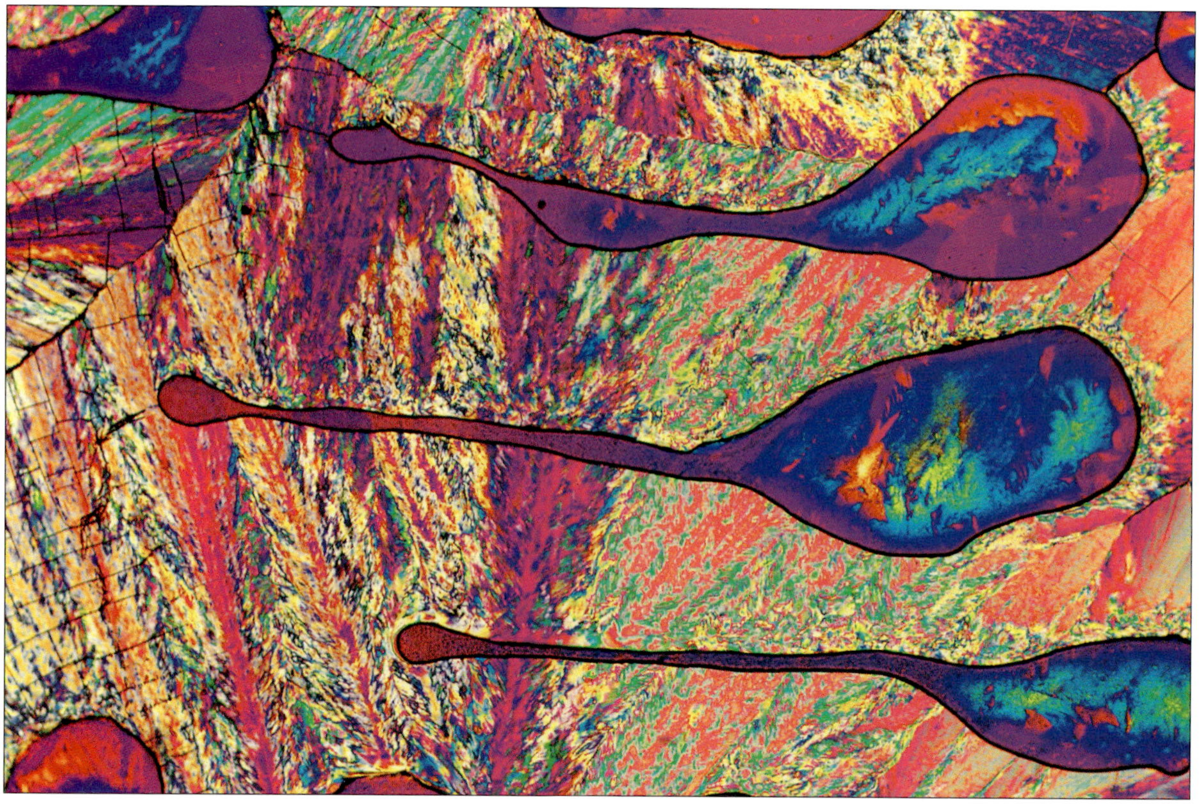

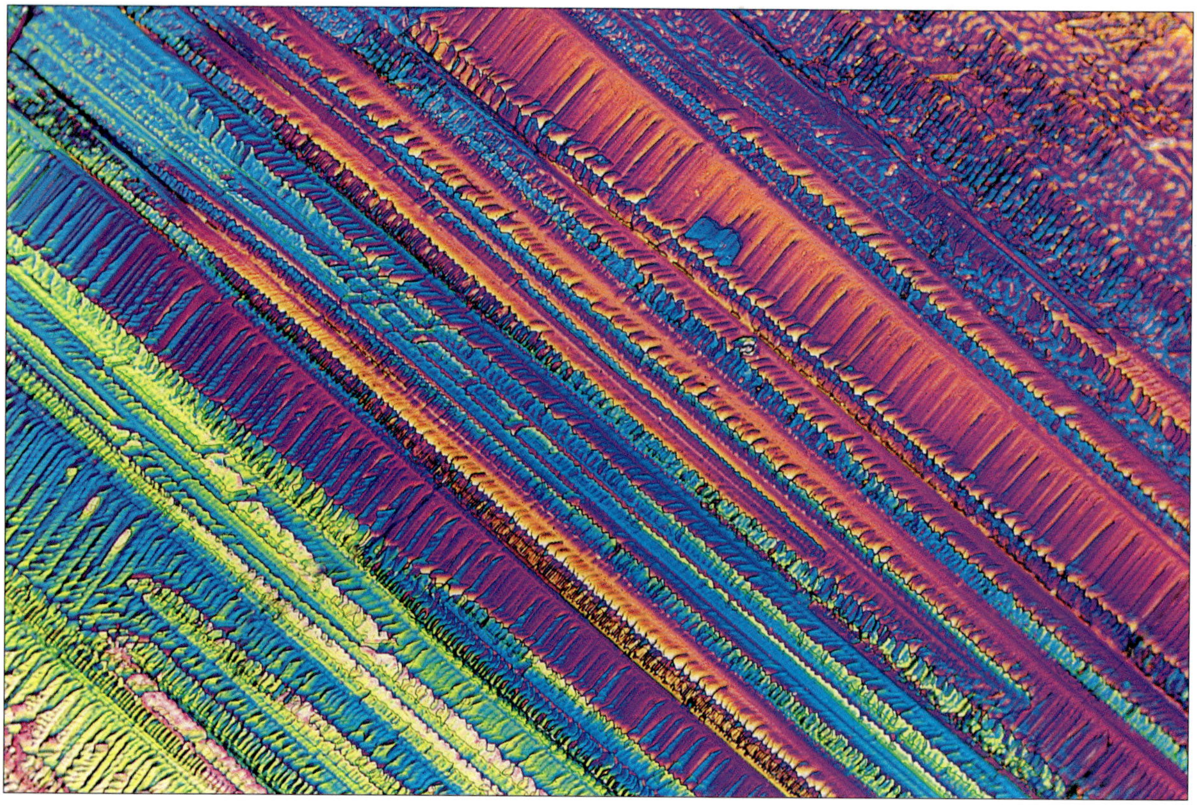

Adenine (Vitamin B4)

Beta Carotene

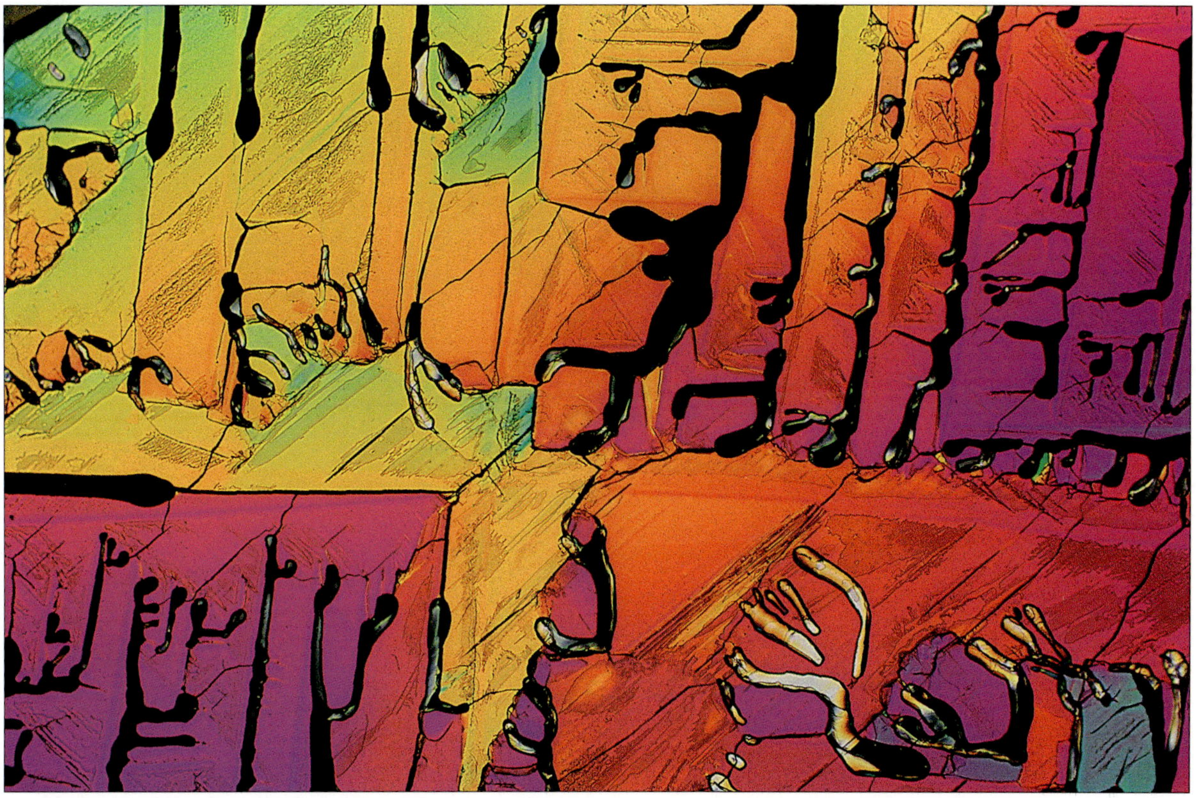

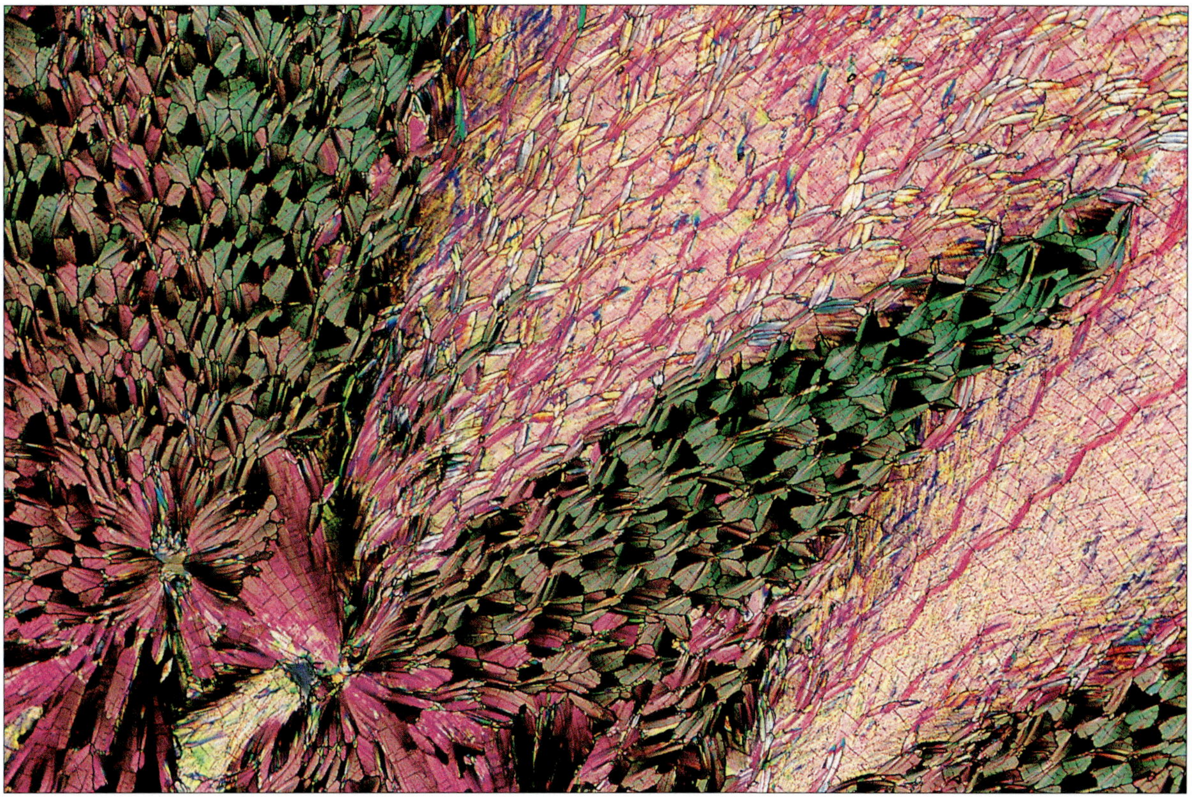

Beta Carotene

Alpha Tocopherol (Vitamin E)

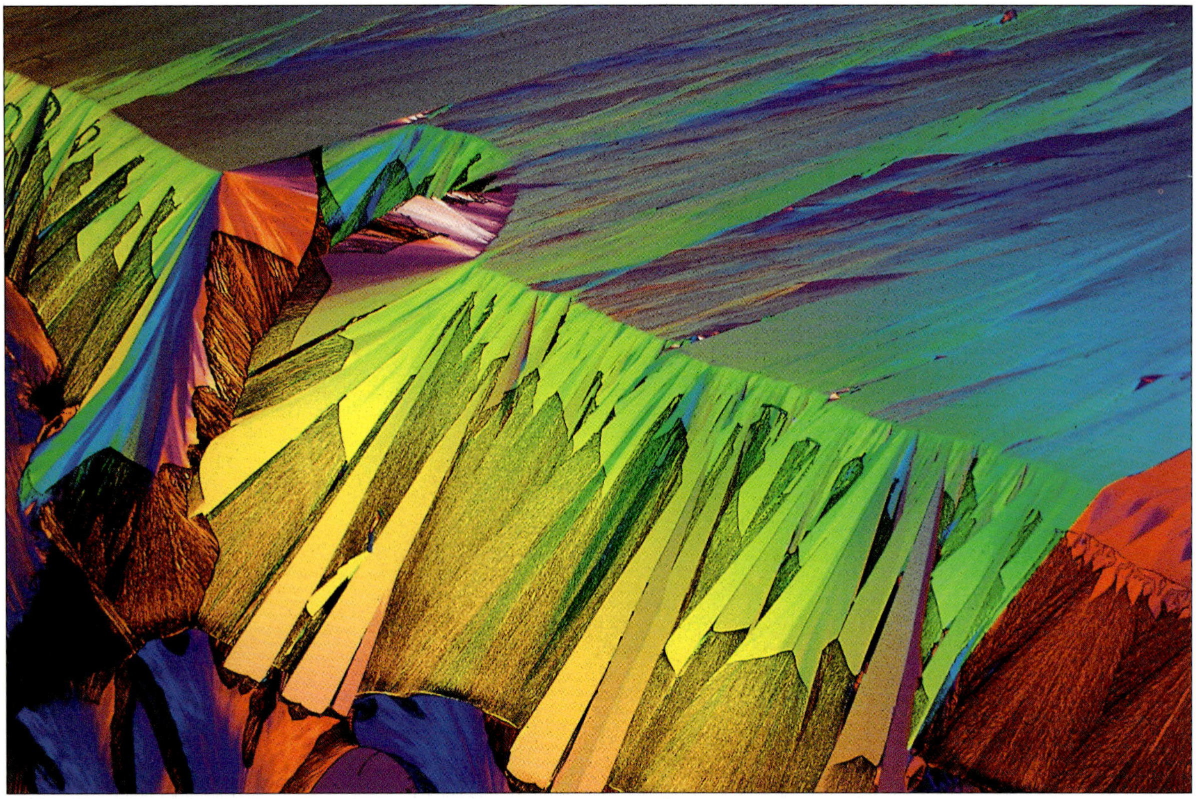

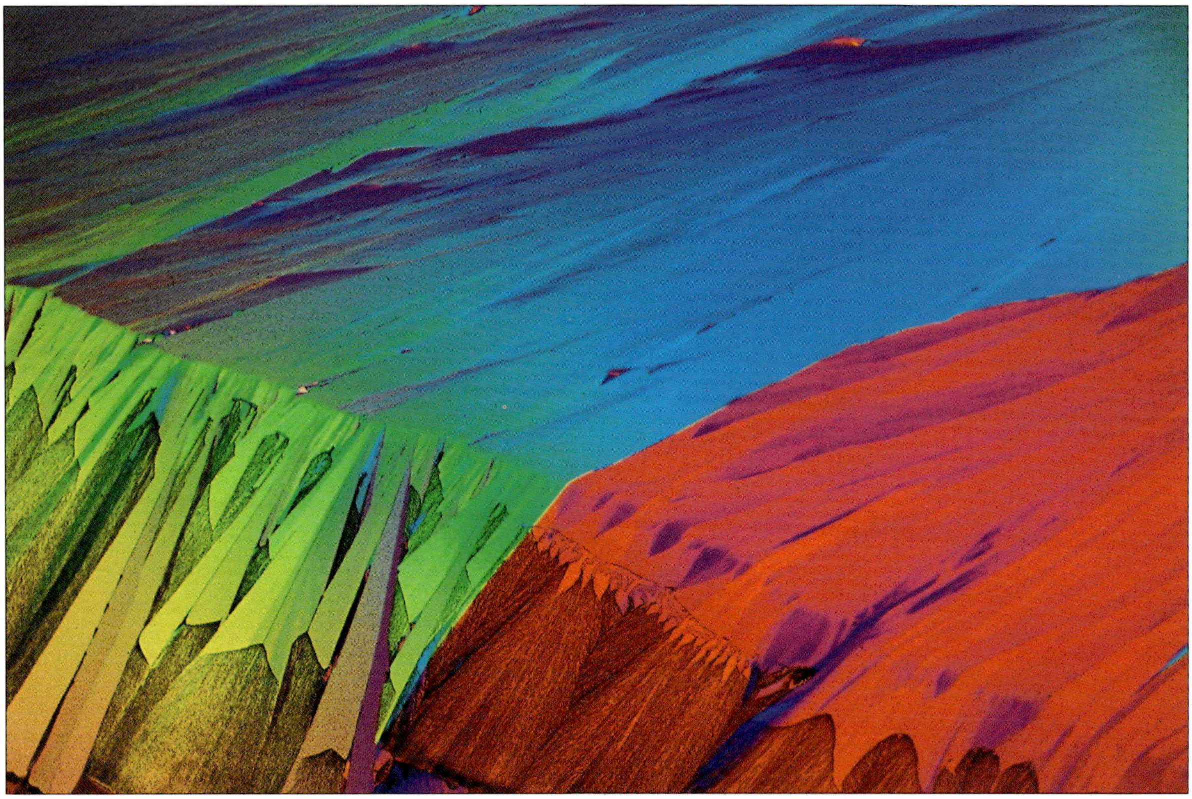

Alpha Tocopherol (Vitamin E)

Riboflavin (Vitamin B2)

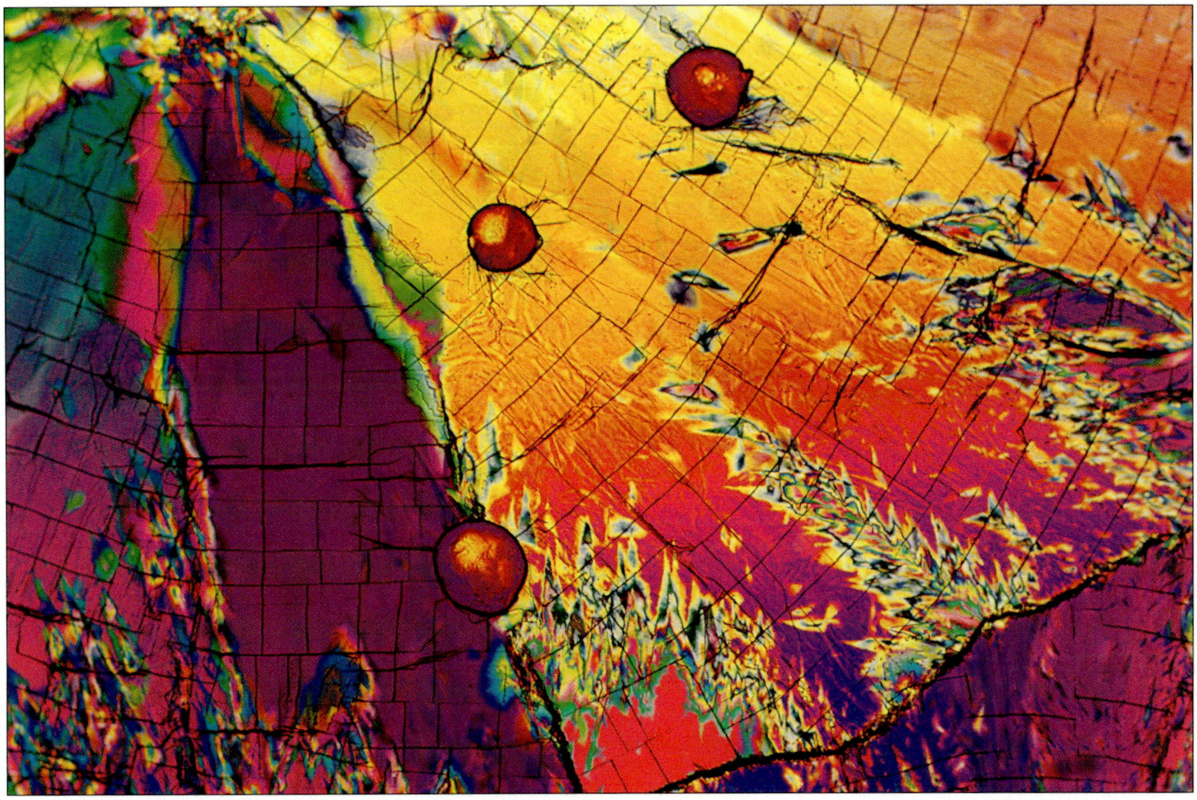

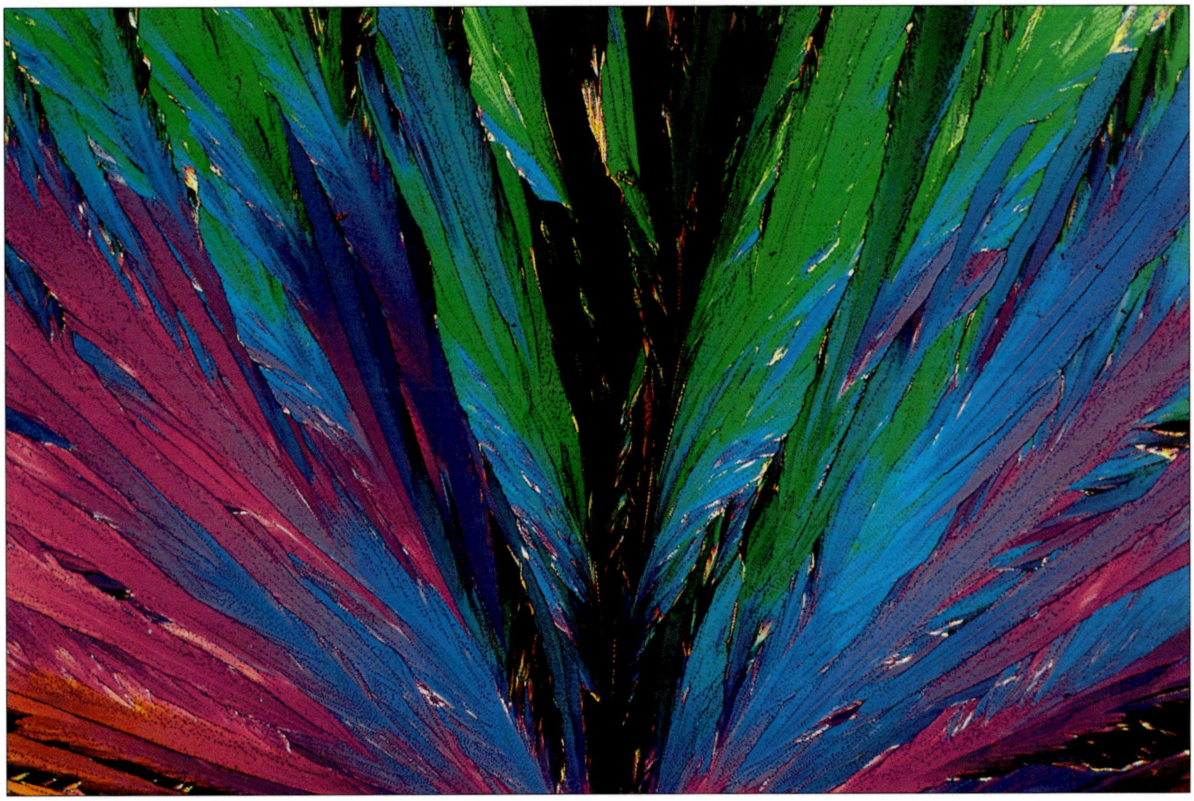

Pyridoxine (Vitamin B6)

Tornadoland

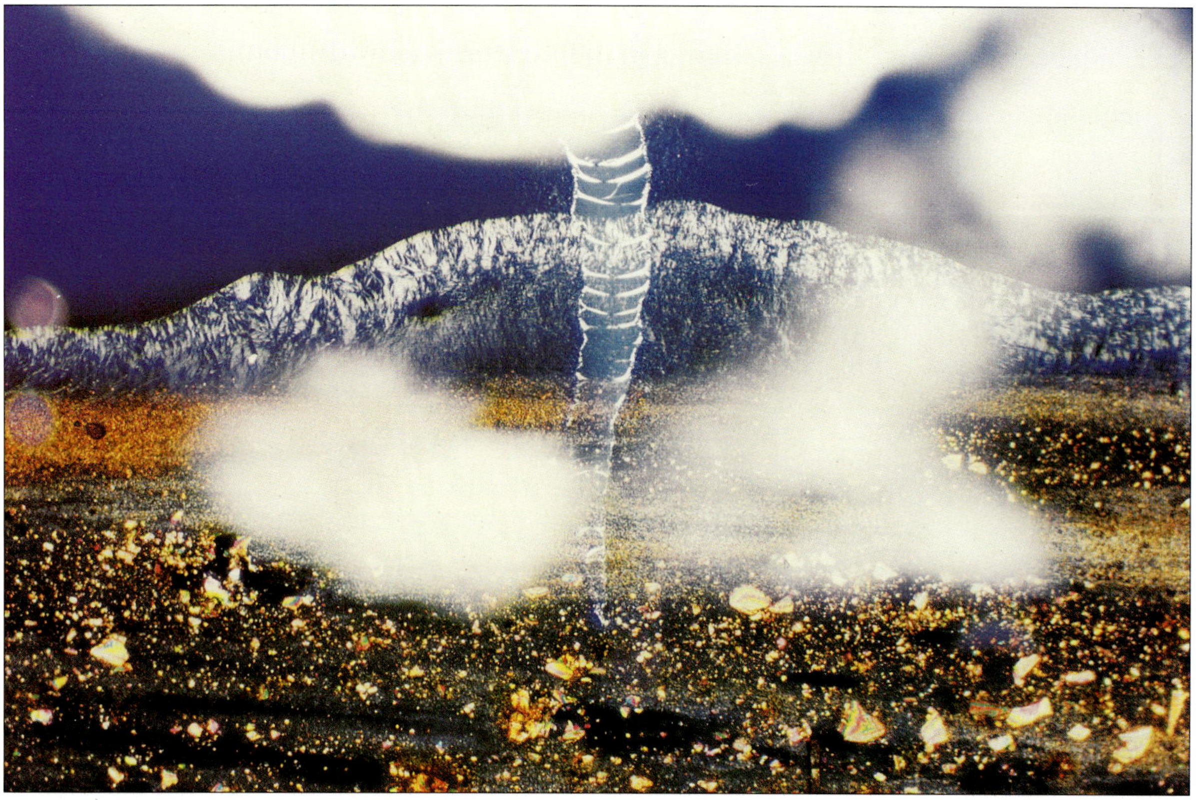

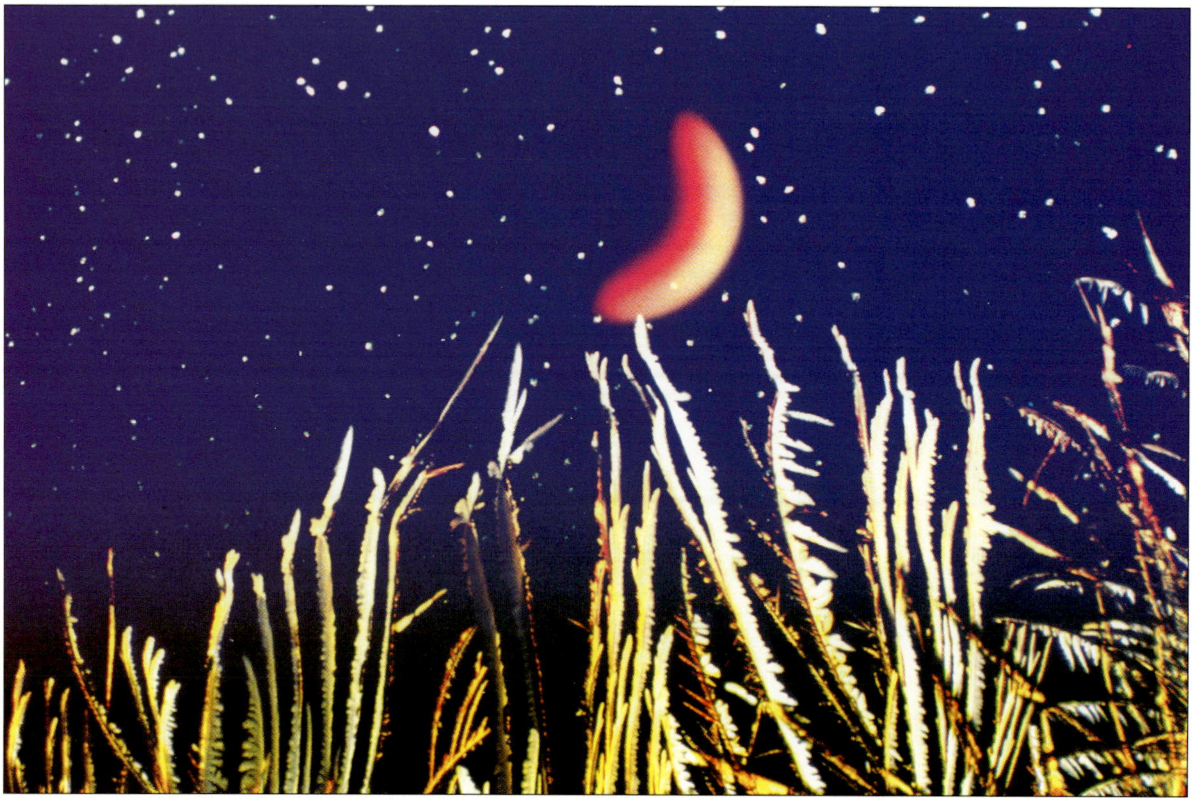

The Stand

Sulfur Canyon

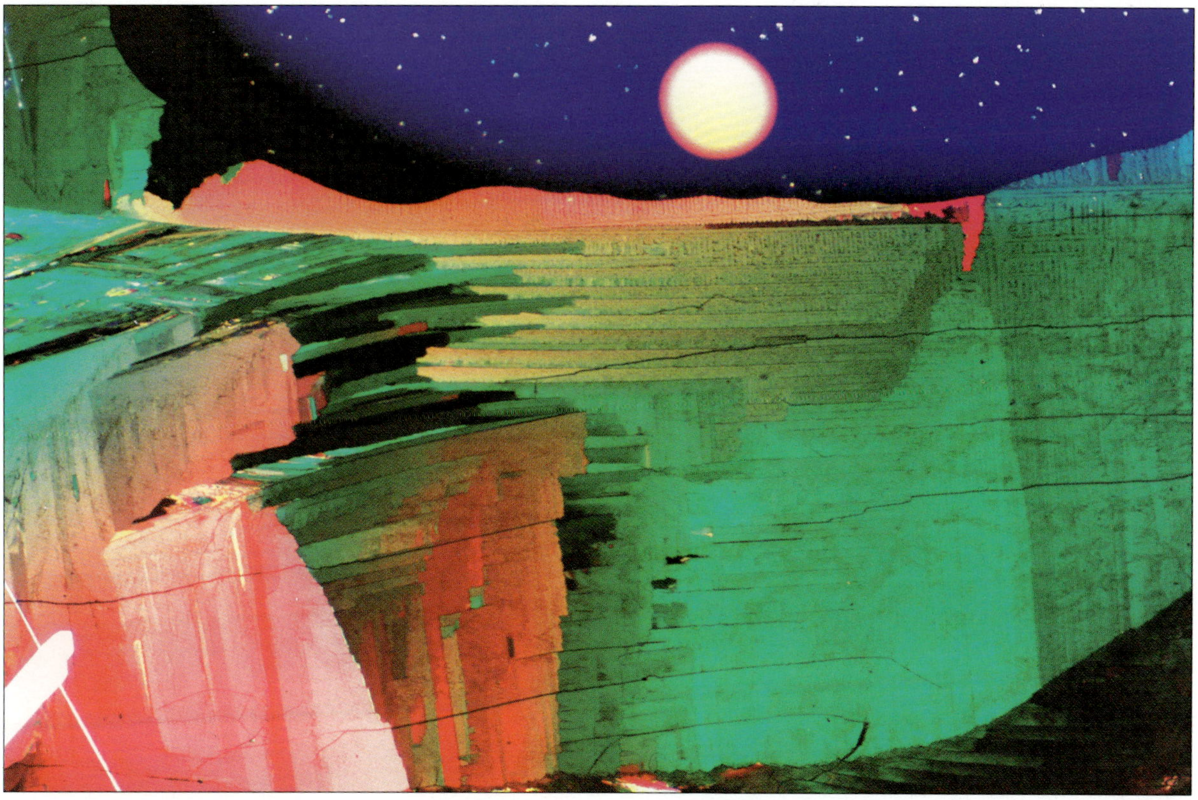

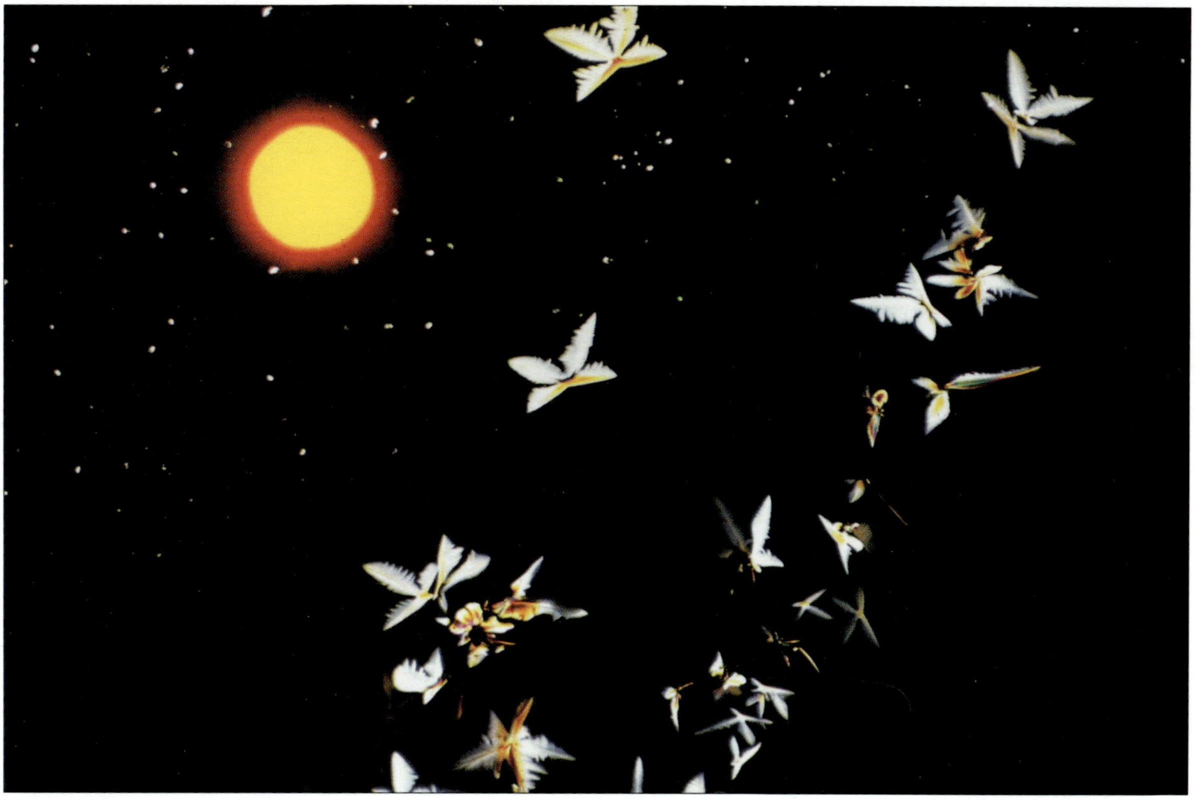

Space Birds

SCIENCE IN IMAGES

When you have many thousands of photographic images, organization becomes a primary consideration in storage, maintenance, and distribution of the collection. I have spent the past few years thinking about how to organize my 40,000+ photomicrographs, and have come up with several solutions. The overall title governing all of my individual portfolios has become ***Science in Images.*** In executing this concept, I have picked one or several of the best images from each daughter collection to be incorporated into the main portfolio. Then, I wrote a nutshell description of the area in science that each collection represents and bundle with the images.

This book is composed of images
from the following collections:

Vitamins,
AIDS Therapeutics,
Liquid Crystalline DNA,
High Temperature Superconductivity,
Steroid Hormones,
Pesticides,
Moon Rocks,
Artificial Sweeteners,
and Buckyballs.

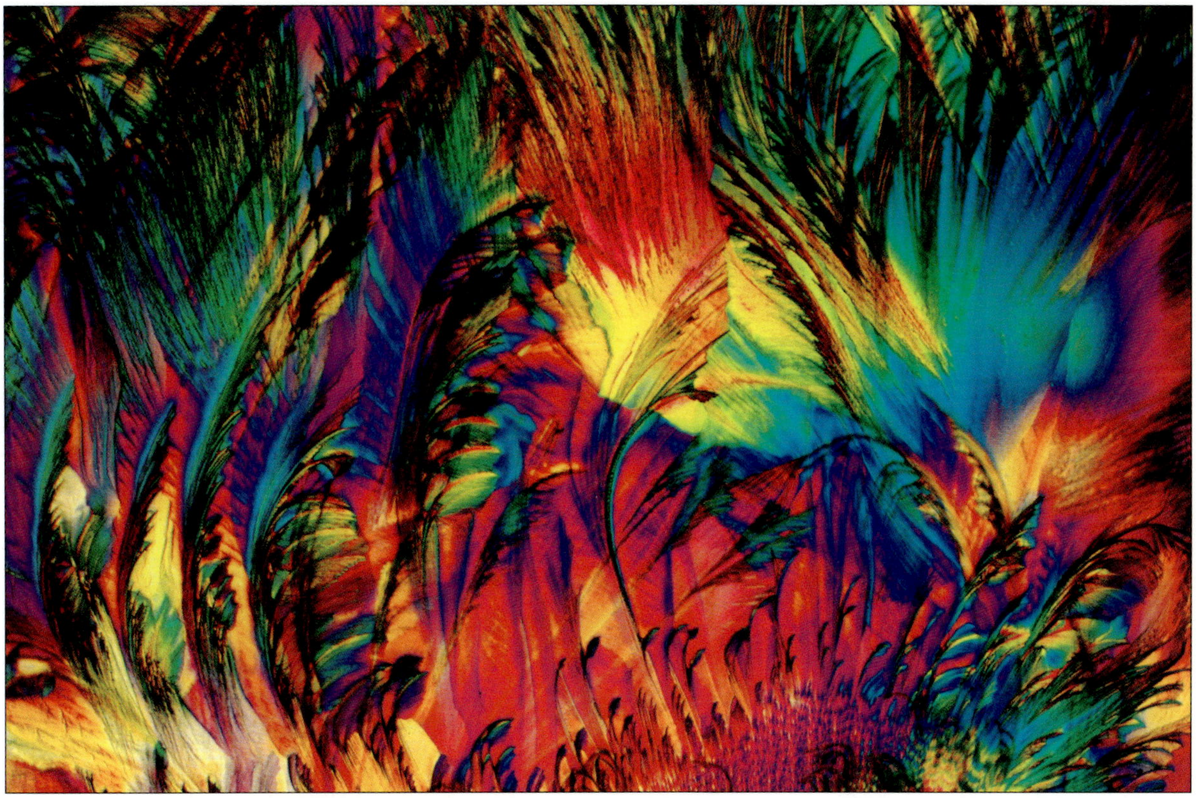

VITAL AMINES

The term vitamin derives from experiments conducted early in this century which indicated that proper nutrition was dependent upon introduction of one or several vital nitrogen-containing amines into the diet. Vitamins are organic molecules (not necessarily amines) that are essential to metabolism in all living organisms. While these molecules serve essentially the same role in all forms of life, higher organisms have lost the ability to synthesize vitamins. There are two major groups of vitamins: the fat-soluble vitamins designated by the letters A, D, E, and K, and the water-soluble vitamins which are referred to as the vitamin B complex. Most vitamins are converted *in vivo* into coenzymes that work with metabolic enzymes to complete their biochemical functions. A lack of proper amounts of vitamins in the diet leads to a host of vitamin-deficiency diseases. The photomicrograph above depicts crystallites of the common vitamin ascorbic acid and has been implicated by Nobel laureate Linus Pauling as a cure for the common cold.

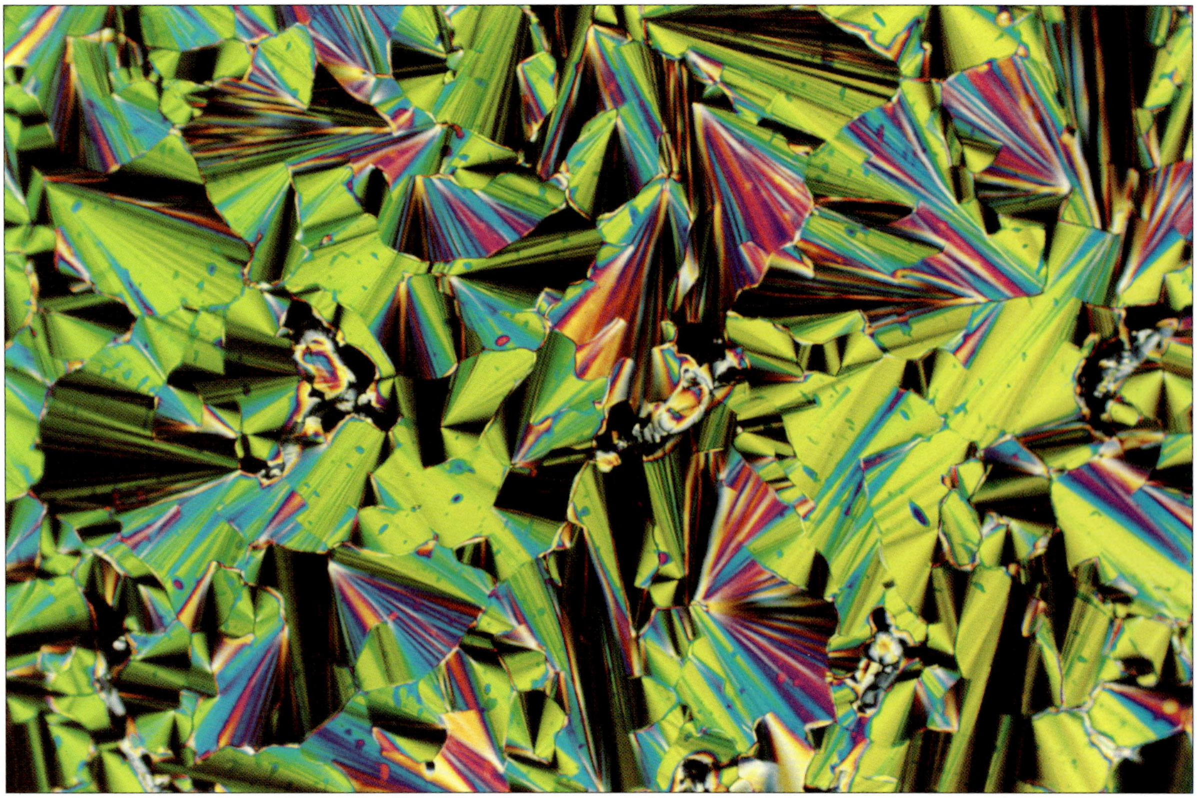

LIQUID CRYSTALLINE DNA

In the laboratory, scientists usually investigate the physical and biological properties of DNA in dilute solution. However, *in vivo* DNA exists in domains where the localized concentrations are very high. As the aqueous solution concentration of DNA is slowly increased, the macromolecule undergoes spontaneous phase transitions in the liquid crystal field. The phase depicted here is a high-density columnar hexatic liquid crystalline phase where the DNA concentration is 400-500 milligrams per milliliter. These are concentrations approaching those found in bacterial nucleoids, dinoflagellate chromosomes, virus capsids, and sperm heads indicating that the DNA in these organelles probably exists in a liquid crystalline state.

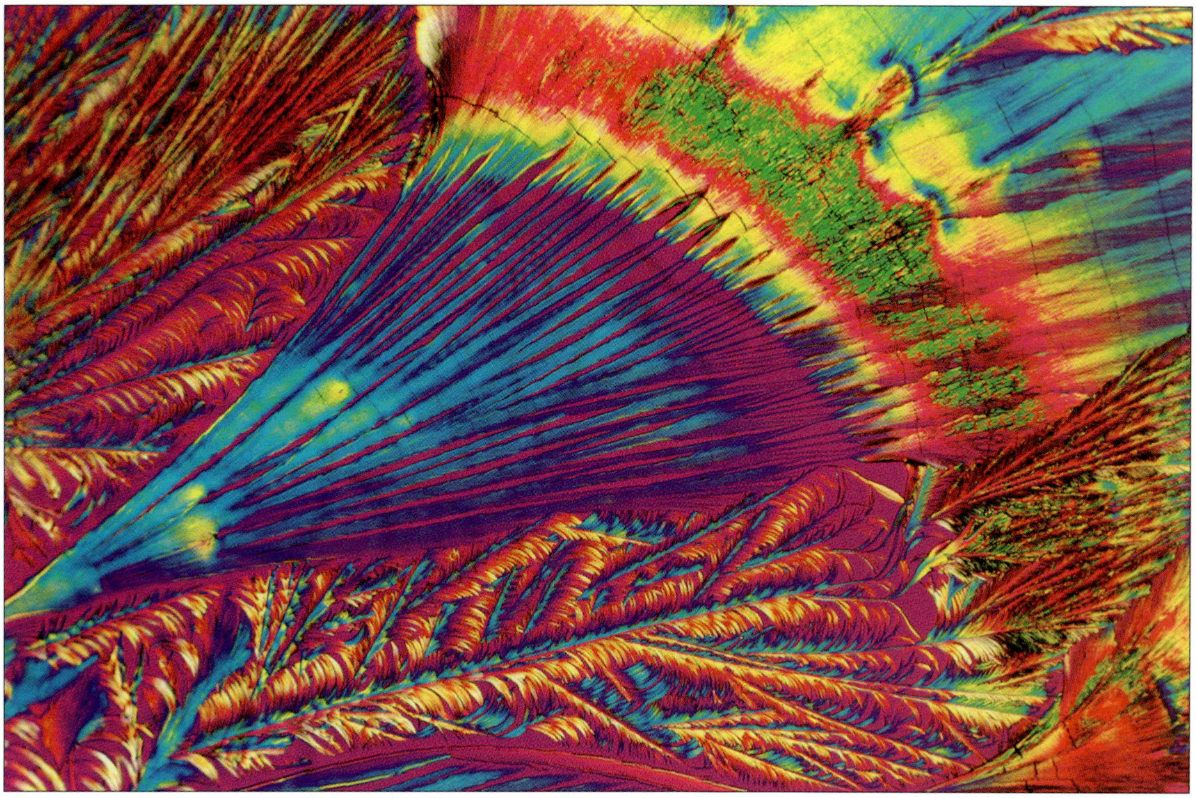

AIDS THERAPEUTICS

It is widely assumed that the Acquired Immuno-Deficiency Syndrome (AIDS) is caused in humans by a retrovirus termed Human Immunodeficiency Virus (HIV). 3'–azidothymidine (also known as AZT, Retrovir® and Zidovudine) is probably the most effective anti-AIDS medication developed to date. During initial clinical trials using a double-blind placebo, the treated study population displayed a fivefold diminution in mortality rates. The drug is phosphorylated *in vivo* to the triphosphate analog which is mistaken for thymidine triphosphate by the viral specific DNA replicating enzyme reverse transcriptase. Subsequently, the azido analog is incorporated into the DNA transcript of viral RNA and causes premature chain termination, thus preventing integration of viral double-stranded DNA into the host genome. This image of AZT crystallites represents a genuine image of hope for a cure from this deadly plague.

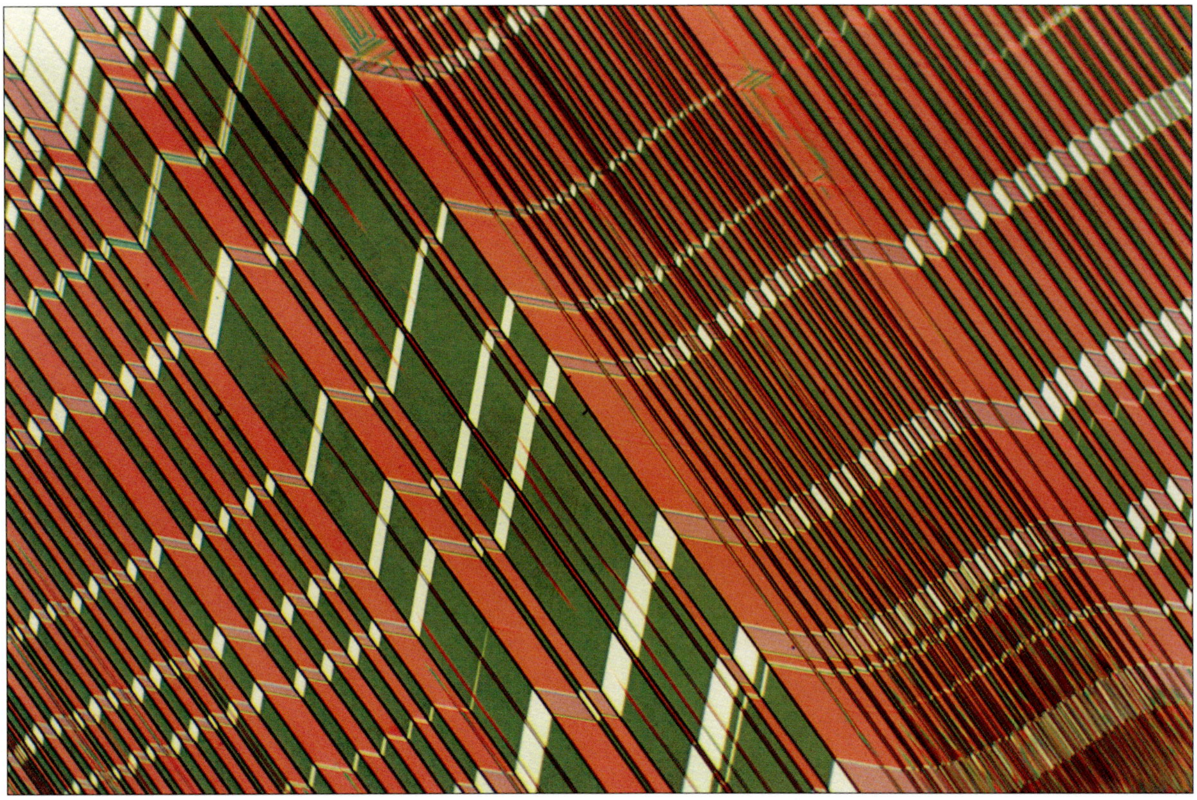

HIGH TEMPERATURE SUPERCONDUCTIVITY

This is a transmitted polarized light micrograph of a thin wafer of the crystalline ceramic material lanthanum aluminate ($LaAlO_3$). This material is termed a *Perovskite* and is a member of a family of ceramics which have recently been demonstrated to superconduct at liquid nitrogen temperatures. The lanthanum aluminate wafer itself does not superconduct but it is a good lattice match for thin film deposition of the superconducting ceramic $YBa_2Cu_3O_{7-x}$ by epitaxy methods such as laser ablation and Metal-Organic Chemical Vapor Deposition (MOCVD). Stair step twinning, very evident in this photomicrograph, interferes with confluent film formation and researchers are hurriedly trying to develop single crystals which are devoid of this twinning.

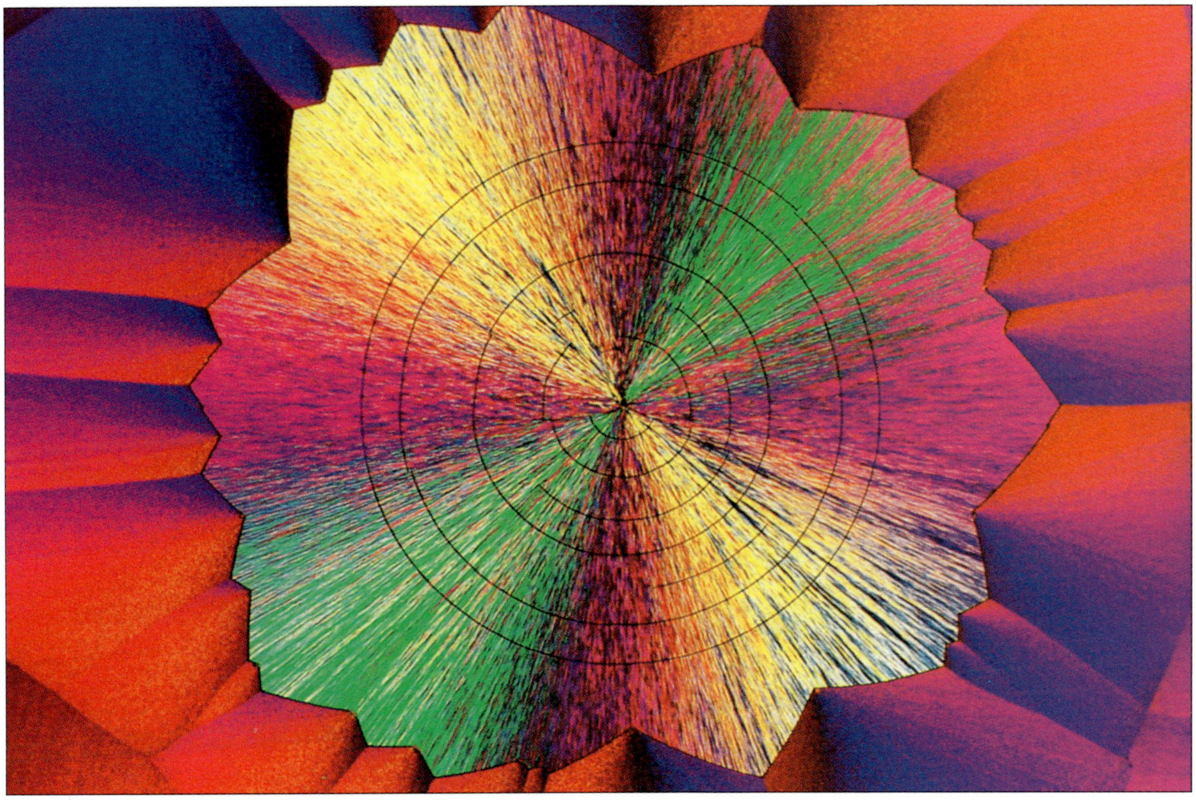

STEROID HORMONES

Cholesterol is the biochemical precursor for the five major classes of steroid hormones: progestagens, glucocorticoids, mineralocorticoids, androgens, and estrogens. The micrograph above illustrating the female steroidal hormone progesterone (a progestagen), is secreted by the ovaries subsequent to ovulation and prepares the lining of the uterus for implantation of an ovum. Progesterone is also essential for the maintenance of pregnancy. The male steroidal hormones are termed androgens, of which the most prominent member is testosterone. Androgens are responsible for the development of male secondary sexual characteristics. The female steroid family of estrogens (such as estradiol and estrone) are required for the development of female secondary sexual characteristics.

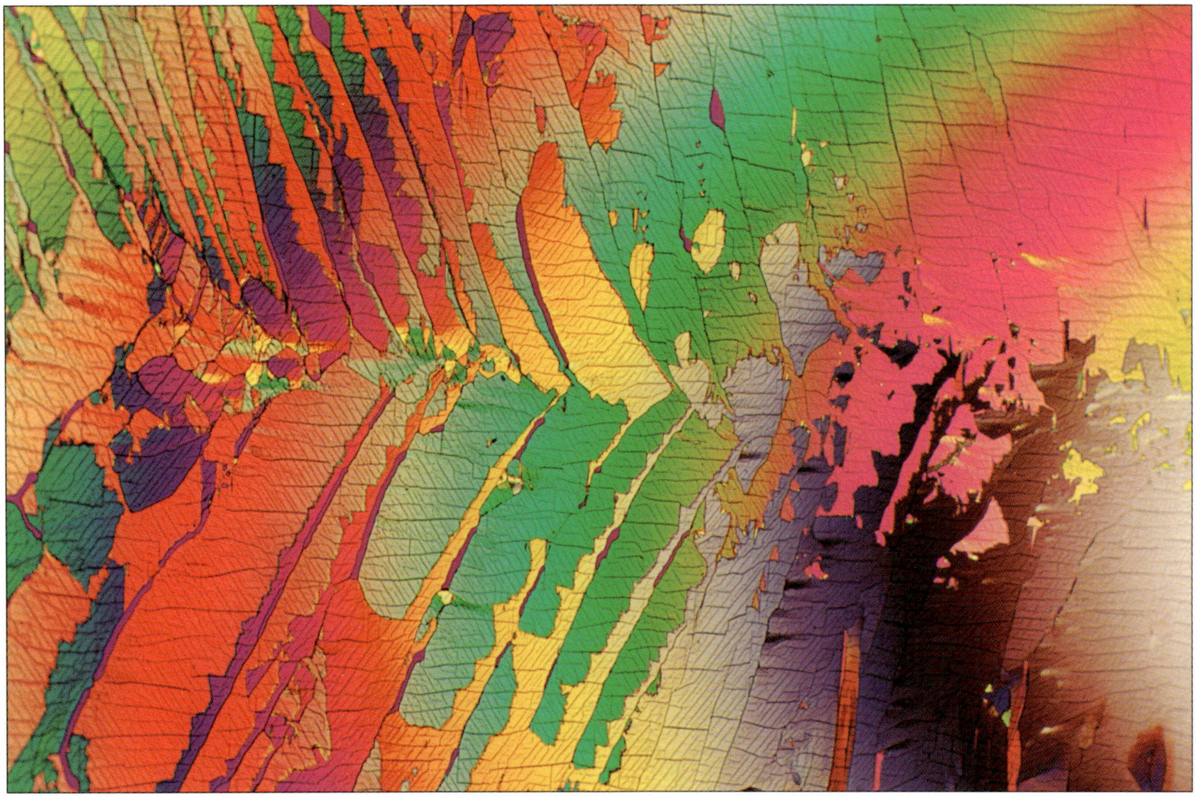

PESTICIDES

The term pesticides derives from the Latin *pestis* for plague and is used to describe plants (weeds), vertebrates, insects, mites, pathogens and other oranisms which occur where we do not want them. With the availability of vast quantities of succulent food and often the absence of natural enemies, pests can reproduce quickly and become a serious agricultural problem. Insects, in particular, often exhibit a dramatic and rapid increase in population size under these conditions. The average soil density of insects is about 9 million per acre with about 10,000 "in flight" above it. The photomicrograph above illustrates crystallites of the broad spectrum insecticide Dichloran, a member of the organochlorine class of pesticides, fathered by the now banned DDT (dichlorodiphenyltrichloroethane). Organochlorines alter both sodium and potassium concentrations in neurons, affecting impulse transmission and causing muscles to twitch spontaneously. They also persist for a long time in the environment.

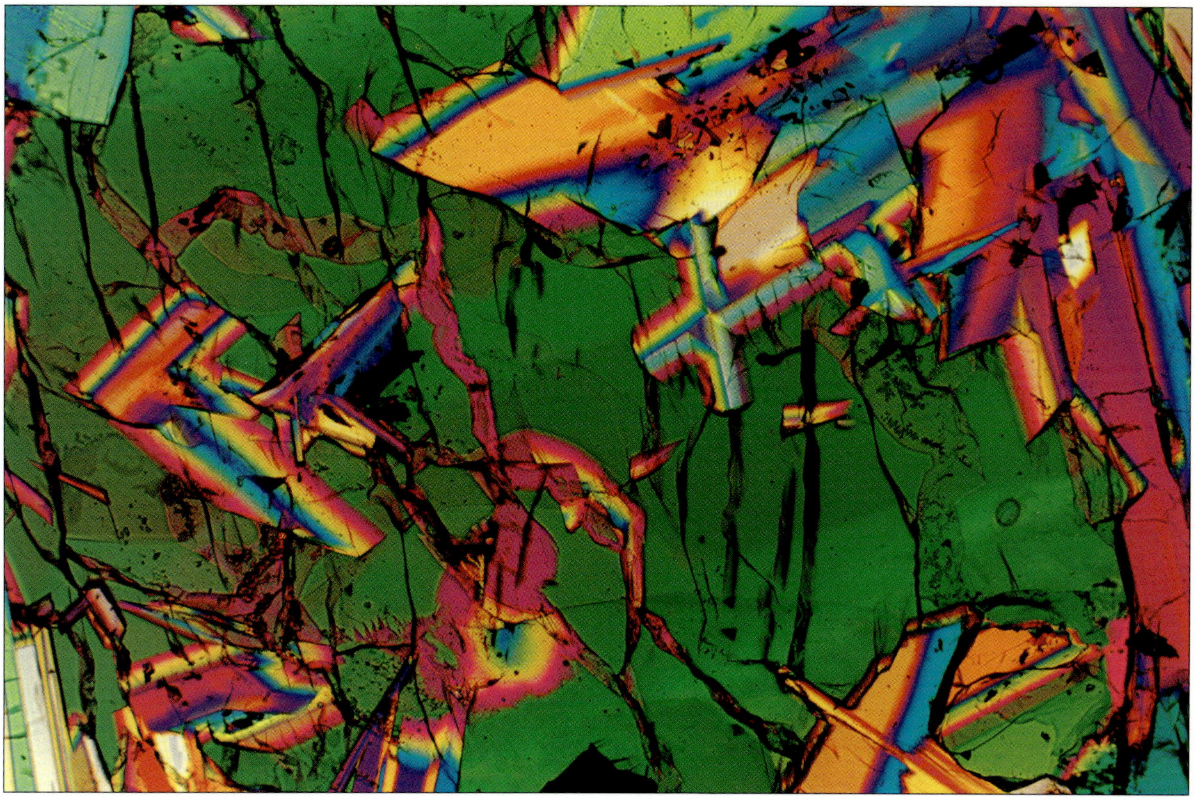

MOON ROCKS

Between 1969 and 1972 the National Aeronautics and Space Administration (NASA) successfully landed 12 astronauts on the lunar surface. The astronauts who visited the Moon carefully collected 2,196 documented samples of lunar soils and rocks weighing a total of 382 kilograms (843 pounds) during approximately 80 hours of exploration. The lunar surface can be roughly divided into two domains. The photomicrograph illustrated here is derived from a polished thin section of a lunar basaltic lava sample collected from the *Oceanus Procellarum* mare region by Apollo 12 astronauts. Mare basalts are volcanic lavas rich in iron and titanium oxide minerals that formed when molten rock from the Moon's interior came to the surface and flowed over large areas.

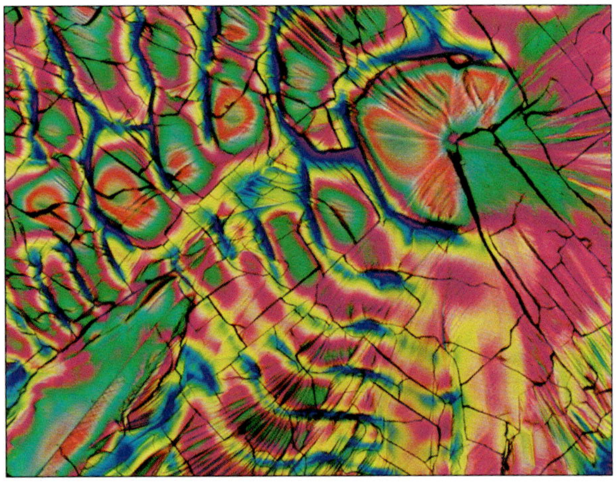

FLAVORS AND FRAGRANCES

Fragrances (perfumes) have been around since the beginning of civilization. They were first used in funeral rites and religious ceremonies, and later in other ways during routine daily life. Early chemists employed a method known as enfleurage, whereby petals of a desired flower are laid on thin layers of purified fats, to extract small quantities of fragrance chemicals from within the petals. The advent of synthetic organic chemistry changed the perfume and flavor industry when aroma chemicals were first isolated and identified in the early 1830s. European chemists identified cinnamic aldehyde from cinnamon oil, and benzaldehyde from bitter almond oil. After 1850, more synthetics became available such as esters of low molecular weight acids and alcohols, methyl salicylate (artificial wintergreen oil) and vanillin. The age of synthetic fragrances really arrived in the 1920s with the introduction of *Chanel No. 5*® perfume, a concocted fragrance that perfumers describe as "aldehydic" in nature. The photomicrograph illustrated on the cover of this collection is a small organic chemical known as *methyl-3-nonenoate*, which when mixed in candy at 6.2 parts per million bestows a flavor reminiscent of watermelon. Naturally, this chemical has been extracted from orris derivatives. Other members of this collection also have natural origins, although they are easily synthesized by organic chemists in the laboratory.

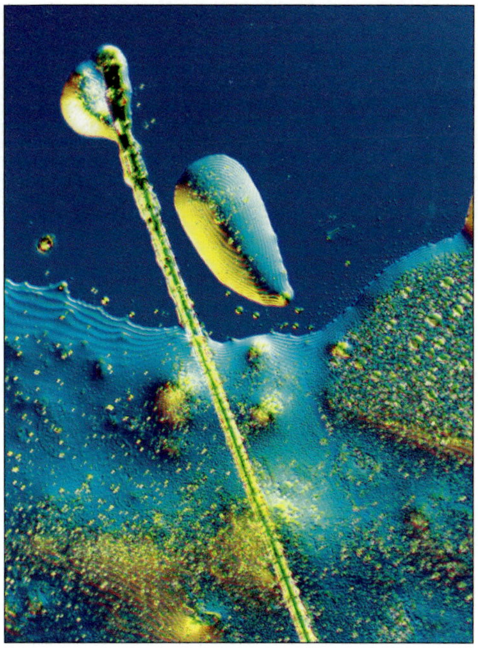

BUCKYBALLS

Buckyballs is an acronym for the 60-carbon alkene buckminsterfullerene, a new form of carbon. The molecule was named after the geodesic dome, invented by Buckminster Fuller, whose structure approximates a truncated icosohedral soccer ball-shaped structure. The fullerenes (C_{60} and C_{70}) are really quite remarkable molecules. They were initially discovered by Richard E. Smalley and his associates at Rice University during experiments aimed at understanding the mechanisms by which long-chain carbon molecules are formed in interstellar space. The molecule has carbon atoms at 60 chemically equivalent vertices that are connected by 32 faces, 12 of which are pentagonal and 20 hexagonal. Buckyballs are very rugged. They can survive collisions with metals and other materials at speeds in excess of 20,000 miles per hour, a speed that would tear most organic molecules apart. Recently, potassium "doped" buckyballs were found to be superconducting at the amazingly high temperature of 18°K. These superconducting buckyballs have the highest critical temperature of any known organic compound and have now been termed "dopeyballs".

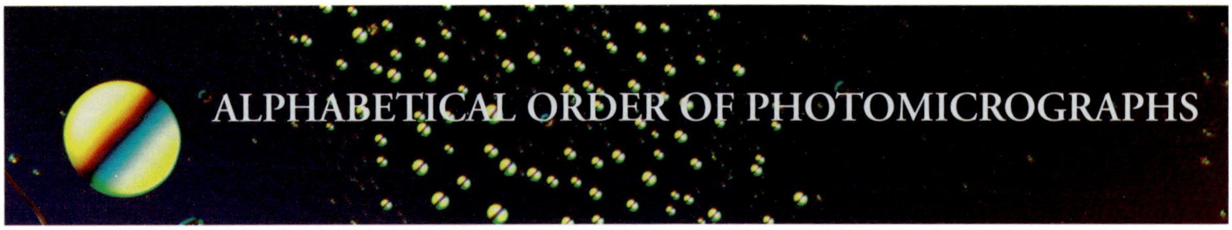

ALPHABETICAL ORDER OF PHOTOMICROGRAPHS

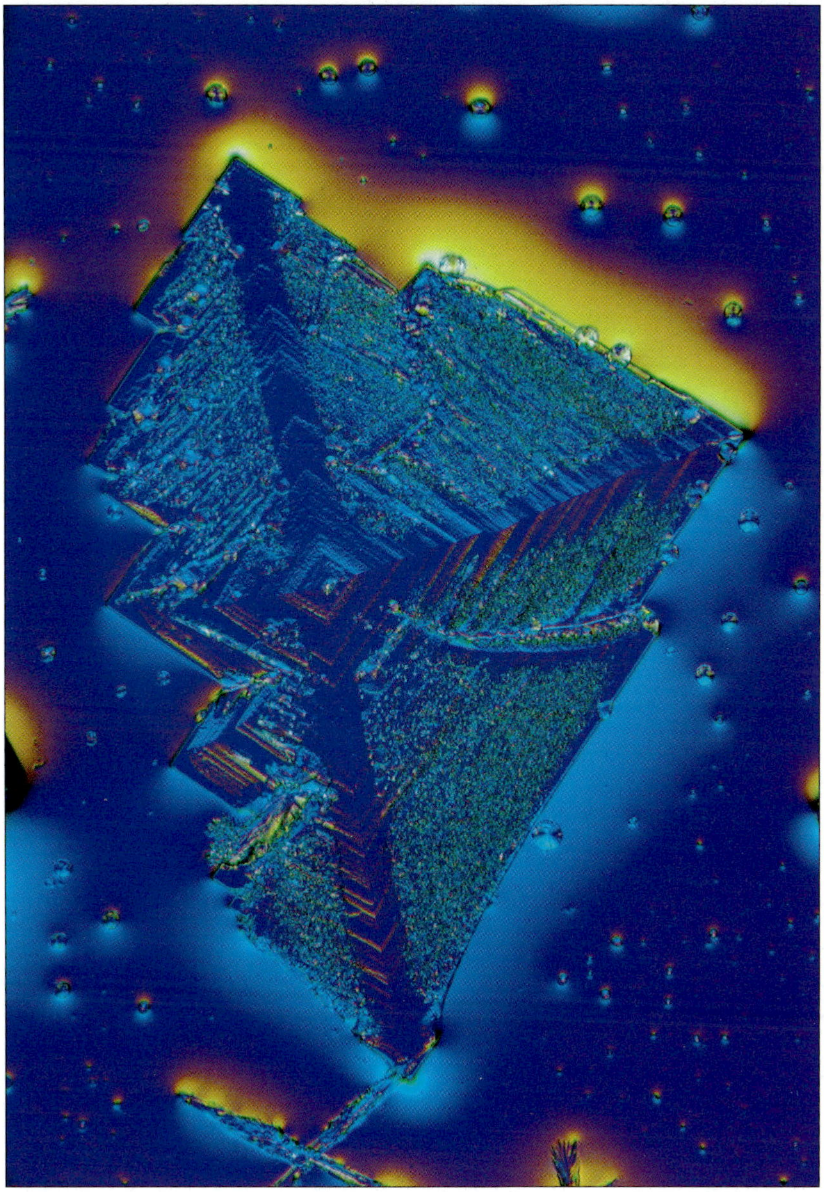

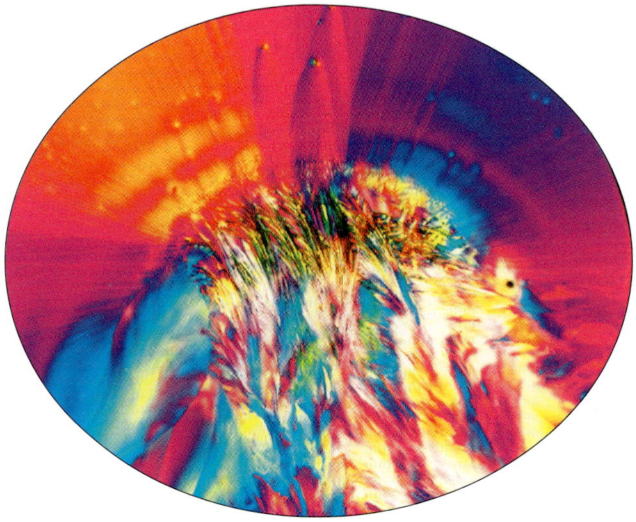